W9-DFX-871

STP 1391

Structural Integrity of Fasteners: Second Volume

Pir M. Toor, editor

ASTM Stock Number: STP 1391

ASTM
100 Barr Harbor Drive
West Conshohocken, PA 19428-2959
Printed in the U.S.A.

Library of Congress Cataloging-in-Publication Data

Structural integrity of fasteners. Pir M. Toor, editor.
    p.cm.—(STP; 1236)
    "Papers presented at the symposium of the same name held in Miami,
Florida on 18 Nov.1992 . . . sponsored by ASTM Committee E-8 on
Fatigue and Fracture" —CIP foreword.
    "ASTM publication code number (PCN) 04-012360-30."
    Includes bibliographical references and index.
    ISBN 0-8031-2017-6
    1. Fasteners. 2. Structural stability. I. Toor, Pir M.
II. ASTM Committee E-8 on Fatigue and Fracture. III. Series: ASTM special
technical publication; 1236.

TJ1320.S77 1995
621.8'8—dc20
ISBN 0-8031-2863-0 (v. 2)   95-12078
                                                                    CIP

Photocopy Rights

Peer Review Policy

Each paper published in this volume was evaluated by two peer reviewers and at least one editor.
The authors addressed all of the reviewers' comments to the satisfaction of both the technical editor(s)
and the ASTM Committee on Publications.
    The quality of the papers in this publication reflects not only the obvious efforts of the authors and
the technical editor(s), but also the work of the peer reviewers. In keeping with long standing publication practices, ASTM maintains the anonymity of the peer reviewers. The ASTM Committee on
Publications acknowledges with appreciation their dedication and contribution of time and effort on
behalf of ASTM.

Printed in Philadelphia, PA
July 2000

Foreword

This publication, *Structural Integrity of Fasteners: Second Volume,* contains papers presented at the Second Symposium on Structural Integrity of Fasteners, held in Seattle, Washington, on May 19, 1999. The sponsor of this event was ASTM Committee E08 on Fatigue and Fracture and its Subcommittee E08.04 on Application. The Symposium Chairman was Pir M. Toor, Bettis Atomic Power Laboratory, (Bechtel Bettis, Inc.) West Mifflin, PA. Those who served as session chairmen were Harold S. Reemsnyder, Homer Research Labs, Bethlehem Steel Corp., Louis Raymond, L. Raymond and Associates, Newport Beach, California, and Jeffrey Bunch, Northrop Grumman Corporation, Pasadena, California.

A Note of Appreciation to Reviewers

The quality of papers that appear in this publication reflects not only the obvious effort of the authors but also the unheralded, though essential, work of the reviewers. This body of technical experts whose dedication, sacrifice of time and effort, and collective wisdom in reviewing the papers must be acknowledged. The quality level of this STP is a direct function of their respected opinions. On behalf of ASTM committee E08, I acknowledge with appreciation their dedication to a higher professional standard.

Pir M. Toor
Technical Program Chairman

Contents

Overview

This book represents the work of several authors at the Second Symposium on Structural Integrity of Fasteners, May 19, 1999, Seattle, Washington. Structural integrity of fasteners includes manufacturing processes, methods and models for predicting crack initiation and propagation, fatigue and fracture experiments, structural integrity analysis and failure analysis. Papers and presentations were focussed to deliver technical information the analyst and designers may find useful for structural integrity of fasteners in the year 2000 and beyond.

The papers contained in this publication represent the commitment of the ASTM subcommittee E08.04 to providing timely and comprehensive information with respect to structural integrity of fasteners. The papers discuss failure approaches, fatigue and fracture analysis techniques, and testing procedures. A current bibliography on matters concerning fastener integrity is included at the end of the technical sessions.

Failure Approaches

The intent of this session was to present failure evaluation techniques to determine the structural integrity of fasteners. Failure mechanisms were discussed in real applications of fasteners from assembly process of a hybrid nylon and steel agricultural wheel to high strength failures in steel components. The primary emphasis was to find the mechanism of failure in the fasteners and to predict the structural integrity.

One of the papers in this session discussed fastener failures in which design inadequacy was identified as a cause of failure. Environmental effects and the accuracy of the loading history were evaluated by reproduction of the failure mode via laboratory simulation. Two possible service conditions that may have contributed to failure were simulated in the laboratory to identify the loading rate and the weakness in the assembly design. Quantitative fractographic methods were used to determine the service loads. The authors concluded that the fatigue stress range and maximum stress can be estimated by quantifying the fracture surface features. The authors suggested that accurate results can be obtained if the tests are conducted using the actual material of the failed studs along with the expected service environment, loading rate, and stress ratio, if these variables are known.

Another paper in this session discussed the life prediction methodologies for fasteners under bending loads. The authors compared the S-N approach with fracture mechanics methodology to predict the bending fatigue life of the fasteners. The authors concluded that the tensile S-N data does not accurately predict the bending fatigue life and the fracture mechanics approach yields a conservative prediction of crack growth.

The last paper in this session discussed the failure analysis of high strength steel army tank recoil mechanism bolts. The bolts failed at the head to shank radius during installation. Optical and electron microscopy of the broken bolts showed black oxide on the fracture surfaces with the characteristic of quench cracks. The crack origin was associated with a heavy black oxide that was formed during the tampering operation. The cause of failure was attributed to pre-existing quench cracks that were not detected by magnetic particle inspection during manufacturing. The author stated that to preclude future failure of bolts, recommendations were made to improve control of manufacturing and inspection procedures.

Fatigue and Fracture

The purpose of this session was to highlight the fatigue crack growth state-of-the-art methodology including testing and analytical techniques. An experimental program to investigate the effect of fasteners on the fatigue life of fiber reinforced composites that are used extensively in the industry discussed the failure mode of these composites. The technical areas where further research is needed were also discussed. Another paper discussed the experimental results of low alloy steel fasteners subjected to simultaneous bending and axial loads. The authors concluded that for a bending to axial load ratio of 2:1, fatigue life is improved compared to axial only fatigue life. The fatigue life improvement was more pronounced at higher cycles than at lower cycles. The authors noted that their conclusions are based on limited data. Another paper in this session discussed the stress intensity factor solutions for cracks in threaded fasteners and discussed the development of a closed-form nondimensional stress intensity factor solution for continuous circumferential cracks in threaded fasteners subjected to remote loading and nut loading. The authors concluded that for $a/D = 0.05$, the nut loaded stress intensity factors were greater than 60% of the stress intensity factors for the remote loaded fasteners.

Analysis Techniques

The intent of this session was to discuss the current analysis techniques used to evaluate the structural integrity of fasteners. The breaking load method, which is a residual strength test, was used in the assessment of stress corrosion in high strength steel fasteners. The authors claim that there is a clear relationship between material, and length of exposure time where SCC is present. The authors concluded that by testing a component rather than a tensile specimen, the effects of materials, machining processes and geometry on SCC resistance on the component can be observed.

Another paper in this session discussed the structural integrity of fasteners by measuring the thread lap behavior using finite element analysis along with the fracture mechanics approach. The author started the discussion by defining, "thread laps," using the fasteners industry definition as a "Surface defect, appearing as a seam, caused by folding over hot metal or sharp corners and then rolling or forging them into the surface but not welding them." The author cited the thread lap inspection criteria in the Aerospace industry as ambiguous and difficult to implement. The author analyzed thread lap using two dimensional, axisymmetric, full nut-bolt-joint geometry finite element models. Elastic-plastic material properties, along with contact elements at the thread interfaces, were used in the analyses. Laps were assumed to propagate as fatigue cracks. The author developed a thread profile with a set of laps and their predicted crack trajectories. It was concluded that laps originating at the major diameter and the non-pressure flank were predicted to behave benignly while the laps originating from the pressure flank are not benign and such laps should not be permitted. An inspection criterion was proposed by superimposing a polygon on the thread. The laps within the polygon would be permissible; laps outside the polygon area would be non-permissible. The author claims that this is a more rational method for the acceptance or rejection of the thread laps.

The last paper in this session discussed some recently developed stress intensity factor solutions for fasteners and their application in NASA/FLAGRO 3.0. The stress intensity factor solutions using a three-dimensional, finite element technique were obtained for cracks originating at the thread roots and fillet radii with a thumb-nail shape. A distinction was made between the rolled and machine cut threads by considering the effect of residual stress. These solutions were coded in the NASA computer code NASGRO V3.0.

Testing Procedures

The first paper in this session discussed the criterion for lifetime acceptance test limits for larger diameter rolled threaded fasteners in accordance with the aerospace tension fatigue acceptance criteria for rolled threads. The intent of this paper was to describe a fatigue lifetime acceptance test criterion for thread rolled fasteners having a diameter greater than 1 in. to assure minimum quality attributes associated with the thread rolling process. The author concluded that the acceptance criterion (fatigue life limit) can be significantly influenced by both fastener and compression nut design features that are not included in aerospace fasteners acceptance criteria.

Another paper in this session discussed an experimental technique to evaluate fatigue crack growth in preflawed bolt shanks under tension loads. The intent of the paper was to discuss the state-of-the-art crack growth testing with respect to applied loads, initial and final crack configuration, and the stress intensity factor correlation. The author concluded that the front of a surface flaw in a round bar can be accurately modeled by assuming a semi-elliptical arc throughout the entire fatigue crack growth process. The author also pointed out that the crack aspect ratio changes during cyclic loading and has a marked influence on the crack propagation characteristics. Therefore, the stress intensity factors in a circular specimen must be determined by accounting for the crack depth to bar diameter ratio and the crack aspect ratio.

The third paper in this session discussed the accelerated, small specimen test method for measuring the fatigue strength in the fracture analysis of fasteners. The method consisted of the use of the rising step load (RSL) profile at a constant R-ratio of 0.1 with the use of four point bend displacement control loading. Crack initiation was measured by a load drop. The application of the procedure was demonstrated by presenting a case history.

Finally, an up-do-date bibliography giving references on stress intensity factor solutions related to fasteners application under axial and bending loading is included for engineering use in determining the structural integrity of fasteners.

Pir M. Toor
Bettis Atomic Power Laboratory
Bechtel Bettis, Inc.
West Mifflin, PA
Technical Program Chairman

Failure Approaches

William Counts,[1] *W. Steven Johnson,*[2] *and Ohchang Jin*[3]

Assessing Life Prediction Methodologies for Fasteners Under Bending Loads

REFERENCE: Counts, W., Johnson, W. S., and Jin, O., "Assessing Life Prediction Methodologies for Fasteners Under Bending Loads," *Structural Integrity of Fasteners: Second Volume, ASTM STP 1391,* P. M. Toor, Ed., American Society for Testing and Materials, West Conshohocken, PA, 2000, pp. 3–15.

ABSTRACT: New polyimide matrix composite materials are leading candidates for aerospace structural applications due to their high strength to weight ratio and excellent mechanical properties at elevated temperatures. The high fatigue resistance of these composites often results in the bolts being the weak link of a structure. Aircraft-quality bolts made of 4340 steel with a minimum UTS = 1241 MPa (180 ksi) were tested in three-point bend fatigue. Two life prediction methodologies were accessed for bending stress: *S*-*N* curves and fracture mechanics. The tensile *S*-*N* curve from the *Mil-Handbook-5* conservatively predicts the bending fatigue life and run-out stress. Crack growth data, in the form of *da*/*dN* versus ΔK, from the *Damage Tolerant Design Handbook* was converted to *a* versus *N* data using five geometric correction factors. None of the five correction factors accurately predict crack growth, but all five correction factors did conservatively predict crack growth.

KEYWORDS: aircraft, aerospace structural applications, aircraft-quality bolts, fatigue resistance

The life of a structure is limited by its weakest link. While there has been a lot of research done on structural materials, the fasteners that hold the structure together have been overlooked. In the aviation industry future supersonic cruise commercial aircraft will be expected to last longer than aircraft of the past. Thermoplastic matrix materials are leading candidates for structural applications due to their high strength to weight ratio and excellent mechanical properties at elevated temperatures. The high fatigue resistance of many polymer matrix composites suggests bolts may be the limiting fatigue factor of composite joints.

Bolt bearing fatigue testing of a structural aerospace composite showed carbon fiber reinforced plastics [CFRP] have a longer fatigue life than the fasteners used to hold them together [*1*]. Testing on a CFRP, as shown in Fig. 1, was carried out in order to determine the bearing fatigue properties of the composite. As the load was increased, no bearing damage was seen in the composite laminate, but the bolts that transferred the load to the composite did begin to fail. The failure of the bolt was not a complete surprise because the bolt was considered slightly undersized. However, the fact that the bolt was undersized does not take away the importance of being able to predict the fatigue life of the bolt.

An understanding of crack growth in bolts will help develop better models, which in turn will better predict and prevent fatigue failures of bolts and the structures they support. The

[1] Graduate student, Georgia Institute of Technology, Woodruff School of Mechanical Engineering.
[2] Professor, Georgia Institute of Technology, Woodruff School of Mechanical Engineering and School of Materials Science and Engineering.
[3] Graduate student, Georgia Institute of Technology, School of Materials Science and Engineering.

3

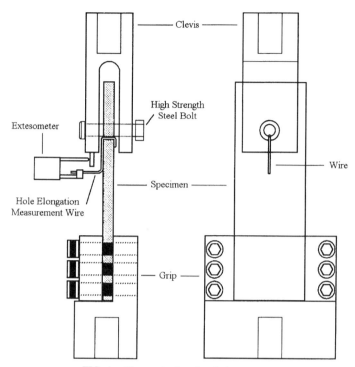

FIG. 1—*Composite bearing fatigue setup.*

two most common life prediction methods are stress life (*S-N*) and fracture mechanics. *S-N* curves are readily available for many materials and are generally generated using tensile loads. *S-N* curves generated using bending loads are uncommon, making it difficult to predict the bending fatigue life. In the absence of any available bending *S-N* data, a published tensile *S-N* curve was compared with the experimentally developed bending *S-N* curve to determine whether tensile *S-N* curves can be used to predict fatigue lives under bending loads. Due to the difference in the stress states, the tensile *S-N* curve predicts a much shorter fatigue life than experimentally observed under bending.

Fracture mechanics has been successful in predicting crack growth in many metals using the following relationship:

$$\Delta K = \Delta\sigma\sqrt{\pi a}F(a)$$

A great advantage of fracture mechanics is it can be applied to many different loading conditions and specimen geometries through the geometric correction factor, $F(a)$. There are numerous geometric correction factors for edge cracks in round bars under bending. These factors vary as the crack gets larger; some factors increase while others decrease [2]. It is unclear how much these variations will affect crack growth predictions and which factor best predicts crack growth.

In order to determine if the correction factors can predict crack growth in bolts under bending, aircraft-quality bolts made of 4340 steel, with a minimum UTS = 1241 MPa (180 ksi), were tested in three-point bend fatigue. Using five geometric correction factors, these experimental data were compared with data from the *Damage Tolerant Design Handbook* [*3*]. While none of the correction factors accurately predict crack growth, they are all conservative.

Materials and Specimens

Specimens for the fatigue tests were aircraft-quality bolts made of 4340 steel, with UTS = 1241 MPa (180 ksi). The bolts were 0.0925 cm (0.375 in.) in diameter and approximately 9 cm (3.5 in.) long. Only 1.6 cm (0.625 in.) of one end of the bolt was threaded, leaving the remaining length smooth.

S-N data for 4340 steel with UTS = 1379 MPa (200 ksi) from the *Mil-Handbook-5* were compared with the experimental *S-N* data because no *S-N* data could be found for 4340 steel with UTS = 1241 MPa (180 ksi). While this difference in UTS will affect the results to some degree, the fact that the bolts had a minimum UTS = 1241 MPa (180 ksi) is somewhat mitigating.

Crack growth data from the *Damage Tolerant Design Handbook* were compared with experimental data. The *da/dN* versus *ΔK* data were taken from plate 4340 steel with UTS = 1241 MPa (180 ksi) tested at 20 Hz. The frequency difference between the experimental crack growth tests run at 10 Hz and the handbook data was not deemed critical since both tests were carried out at relatively high frequencies in a dry air environment.

Testing Techniques

The 4340 bolts were tested in three-point bend fatigue using a servo-hydraulic test frame. The three-point bend fixture was chosen because it best simulates a fastener in a double shear application. All fatigue tests were run at room temperature, a frequency of 10 Hz, and an R-ratio of 0.1. A number of fasteners were cycled at various loads until they failed and an *S-N* curve was developed. Run-out for these tests was one million cycles.

Crack growth tests were also run in three-point bend at 1427 MPa (206 ksi) and 1449 MPa (210 ksi) at 10 Hz and an R-ratio of 0.1. The region where the crack was expected to initiate and grow was polished to provide a smooth surface on which the replicant could be taken. The crack growth was measured using acetate replicants on the surface of the bolts, thus measuring the surface crack length. The acetate replicants provided an excellent copy of the crack, which were viewed and measured under an optical microscope.

Analysis

Bending Stress

The bending stress of the bolt was calculated using beam theory.

$$\sigma = \frac{My}{I} = \frac{32M}{\pi d^3} \qquad (2)$$

where

M = bending moment
y = distance from neutral axis
I = moment of inertia
d = diameter

For specimens under three-point bend, the bending moment (M) is equal to the following

$$M = \frac{\text{Load}}{2} \cdot \frac{\text{Span}}{2}$$

Geometric Correction Factors [F(a)]

Exact solutions for surface cracks in rods under bending are not available because of the complexities of the problem [2]. For example, the stress intensity factor varies along the crack and the crack shape changes as it grows. Therefore, varying assumptions are made to simplify the problem. These assumptions lead to a number of different correction factors for unnotched round bars under bending. A comparison of various correction factors is not easy due to the differing assumptions on which they are based. The easiest criterion on which to compare the correction factors is crack shape: straight crack or semi-elliptical crack. For this research, the crack shape was observed to be semi-elliptical and thus correction factors that assume a straight crack were ignored. The geometry of the specimen is shown in Fig. 2. In this case, the maximum K value was assumed to be at "A." The following five correction factor solutions are considered: (1) Daoud [5], (2) Forman [6], (3) Newman [7], (4) Carpinteri [8], and (5) Murakami/Tsuru (*Stress Intensity Factors Handbook*) [9].

There are two different methods by which the geometric correction factor was determined: finite-element method and manipulating existing similar solutions to fit new conditions. The finite-element method was used by Carpinteri, Daoud, and Newman. Despite being derived in a similar fashion, the results are quite different. Daoud used a two-dimensional plane-stress finite-element method to calculate the strain energy release rate [5]. The strain energy release rate is determined as the rate of change of elastic energy in the bar for successive positions of the circular arc front. These values are comparable to results from a three-dimensional analysis. Daoud determined the normalized strain energy release rate, which was equal to F:

$$\left(F = \left[\frac{\sqrt{EG}}{S\sqrt{\pi a}} \right] \right)$$

Newman used a nodal-force finite-element method for a wide range of nearly semi-elliptical

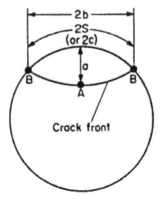

FIG. 2—*Crack shape and geometry.*

surface cracks [7]. He assumed the crack fronts intersect the free surfaces at right angles. Carpinteri also used a three-dimensional finite-element analysis to determine the correction factor [8]. The finite-element model determined the stress field due to a bending load and this stress field was used to calculate the correction factor. Carpinteri's correction factor predicts higher stress intensity at the surface (B) rather than the interior (A). This surface correction factor was used as a comparison to the others.

The second method for determining the correction factor is by taking existing solutions that are similar and manipulating them to fit the new conditions or by fitting existing data to the current problem. Forman used the latter technique to derive a correction factor for round bars under bending [6]. The Forman correction factor was derived from rectangular bar solutions from Tada, which were then multiplied by the factor $(1.03/1.12)(2/\pi)$ to agree with results of Smith for a circular arc front. Murakami and Tsuru derived a different correction factor for bending using existing solutions [9]. They assumed that the ratio of the stress intensity factors for tension and bending in two dimensions is equal to the same ratio in three dimensions and, therefore, the ratio of the correction factors would be the same $(K^B_{I\,3}/K^T_{I\,3} = K^B_{I\,2}/K^T_{I\,2})$. Since three of the correction factors are already known, the fourth $F^B_{I\,3}$ can be found:

$$(F^B_{I\,3} = F^T_{I\,3}\,[F^B_{I\,2}/F^T_{I\,2}])$$

Other correction factors for semi-elliptical cracks, such as the one by Athanassiadis, were not used because they did not contain enough information on the crack shape observed in this case [10]. All of the correction factors used for this research are shown in Fig. 3.

Daoud, Forman, and Murakami provide equations to calculate the correction factors. Newman and Carpinteri provide tabulated data rather than equations. A fourth-order polynomial function was fit to the tabulated data to calculate the correction factors.

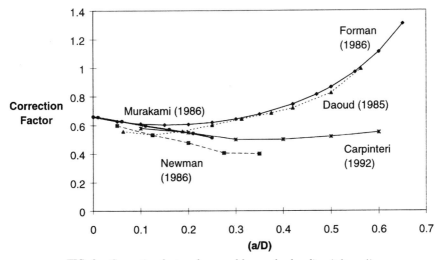

FIG. 3—*Correction factors for round bar under bending* (a/c = 1).

Results and Discussion

Fatigue S-N Approach

Most of the modern S-N data in the literature were developed using tensile loads, with little to no S-N curves for bending loads. We were unable to find bending S-N curves for 4340 because, under bending, the stress in the outer ligament, where the crack initiates and grows, is a tensile stress. Perhaps, tensile S-N data could be used to predict the cycles to failure in this tensile stress region.

A tensile S-N curve taken from *Mil-Handbook-5* for a similar 4340 steel with UTS = 1379 MPa (200 ksi) was compared with the experimental bending S-N curve [4]. The tensile S-N curve did not extend to the high stress levels observed in the bending test. Therefore, for a comparison to be made, the tensile S-N curve was extrapolated to the higher bending stresses using the equivalent stress equation given in the *Mil-Handbook-5* for this particular steel. The equivalent stress equation took into account R-ratio and maximum stress. The comparison of the extrapolated and experimental curves can be seen in Fig. 4.

Unfortunately, the two S-N curves do not overlap. The bending curve is shifted to the right of the tensile curve and has a run out (1 million cycles) stress of 1379 MPa (200 ksi) compared to that of the tensile curve of 689 MPa (100 ksi). The higher stress levels and the higher run-out stress observed in the bending tests are due to the difference in the bending and tensile stress states. The maximum stress under bending is observed only in the outer ligament of the bolt. In tension, the maximum stress is observed throughout the cross section of the bolt. Because a smaller area of the bolt experiences the maximum load under bending, there is a smaller probability of a critically sized flaw being present in the high stress area. Thus, the bolt can go to higher stress levels and will have a higher run-out stress under bending.

The bending S-N curve predicts lives that are approximately ten times greater than those from the tensile S-N curve and run-out stresses that are twice as high. Even though the outer ligament stress under bending is tensile, the tensile S-N curves do not provide an accurate

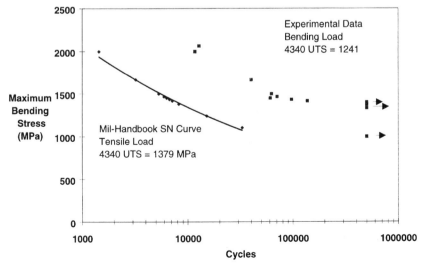

FIG. 4—*Handbook* S-N *curve and experimental* S-N *curve.*

estimation of the bending fatigue life. Therefore, tensile S-N curves should not be used to predict bending fatigue lives.

Fracture Mechanics Approach

Crack Shape—Because it is assumed that the maximum value of K occurs at point "A" (see Fig. 2), the crack length "a" is important. While measuring the crack length "a" is not done very easily, measuring the crack length "c" is easy. However, the crack length "c" is not very useful since the correction factors depend on "a." Therefore, in order to relate the easily measurable crack length "c" to the important crack length "a," the crack shape for 4340 bolt was determined.

To determine the crack shape, four bolts were precracked to various lengths at 1448 MPa (210 ksi) and placed in an oven at 300°C (570°F) for 1 to 3 h. After this time, the exposed surfaces of the metal, including the crack, turned either blue or gold. The bolt was quenched in water and placed in liquid nitrogen to further embrittle the metal. The frozen bolt was placed in the three-point bend fixture and broken with a one-time maximum load. The exact crack shape was then easily discernible from the rest of the metal. An approximation of the final crack shape was estimated from failed specimens. On the failed specimens, the area of crack growth was a lighter gray than the area of fast fracture. However, it was difficult to distinguish the exact transition point from crack growth to fast fracture, making the final crack shape measurement more of an approximation.

The results of the crack shape tests are shown in Fig. 5. The smallest crack measured, $c = 0.29$ mm (0.011 in.), had a circular shape ($a/c = 1$). Three other tests verified the crack shape remained circular through $c = 2.1$ mm (0.08 in.). In crack growth experiments, the last measured surface crack length prior to failure was 2.7 mm (0.11 in.), which failed within 500 cycles. Therefore, it is reasonable to assume that Fig. 5 shows the crack shape through much of the fatigue life of the bolt. The final crack shape was not circular but semi-elliptical [$(a/c) = 0.7$]. This flattening of the crack front appears to happen in the late stages of crack growth.

The crack shape remaining circular was quite unexpected. A possible explanation for this is the surface is in a state of plane stress, implying that there would be more resistance to cracking. However, this part of the crack is also farther from the neutral axis, meaning the stress is higher at this point. The interior is in a state of plane strain, less resistance to cracking, but is closer to the neutral axis, lower stress. These competing mechanisms then cancel each other out, allowing the crack to grow with a constant shape. Athanassiadis predicts $a/b = 0.78$ [*10*]. The final observed crack shape was at $a/c = 0.7$. It appears the crack shape variation from circular to semi-elliptical occurs in the last stages of crack growth, stabilizing at the previously predicted a/b ratios. Although an unexpected result, the research shows the crack shape does not change during initiation and stable growth.

Crack Growth Data

Crack growth of 4340 bolts under bending was studied to predict a fatigue life using fracture mechanics. The maximum observed life before run-out (1 million cycles) was 137 221 cycles at 1413 MPa (205 ksi). This early run-out limited the stress levels available for crack growth. The two stress levels chosen were 1448 MPa (210 ksi), at which there were approximately 80 000 cycles to failure, and 1427 MPa (207 ksi), at which there were approximately 50 000 cycles to failure.

FIG. 5—Crack shape (σ_B = 1148 MPa, Freq. = 10 Hz).

The results of the crack growth test are shown in Fig. 6. At 1448 MPa (207 ksi), the smallest crack measured was approximately 0.16 mm (0.006 in.) and the largest crack measured before failure was 2.7 mm (0.11 in.) with failure occurring within 500 cycles. At 1427 MPa (207 ksi), the smallest crack measured was 0.10 mm (0.004 in.) and the largest crack measured before failure was 2.9 mm (0.11 in.), with failure occurring within 70 cycles. The final crack lengths, determined from the fracture surfaces, were both approximately 4 mm (0.16 in.). It appears the crack grows rapidly during the final cycles.

The experimental a versus N curve was converted to an a versus da/dN curve. This curve shows the three typical stages of crack growth, similar to that seen in tension tests of both approximately 4 mm (0.16 in.). It appears the crack grows rapidly during the final cycles.

The experimental a versus N curve was converted to an a versus da/dN curve. This curve shows the three typical stages of crack growth, similar to that seen in tension tests of other specimen geometries. Both the a versus N and a versus da/dN curves show reasonable crack growth data which can be used with fracture mechanics to predict the crack growth.

Crack Growth Predictions

As discussed earlier, a major difficulty in applying fracture mechanics to round bars is the numerous correction factors available in order to calculate K. To evaluate which correction factor best predicts crack growth in round bars, data from the *Damage Tolerant Design Handbook* were compared with the experimental data [*3*]. The handbook da/dN versus ΔK curve was converted to an a versus N curve for both experimental stress levels using the various correction factors.

The first step of the conversion was plotting the handbook da/dN versus ΔK data and determining the Paris Law constants C and m. The Paris Law equation, $da/dN = C\Delta K^m$, was integrated to determine the corresponding a versus N-curve. This integration is shown below in

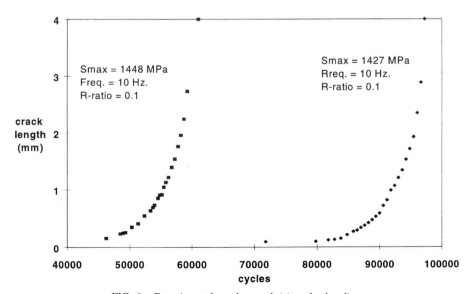

FIG. 6—*Experimental crack growth* (c) *under bending.*

$$\int_{ai}^{af} \frac{1}{C(\Delta\sigma\sqrt{\pi})^m (\sqrt{a}F(a))^m} = N \qquad (4)$$

where a_f = final crack size, a_i = initial crack size, C and m = Paris Law constants, a = crack size, $\Delta\sigma$ = stress range, and $F(a)$ = geometric correction factor.

Since the correction factor was a function of crack length, the integral was determined numerically. The initial crack size (a_i) was taken from the experimental da/dN versus a curves as the crack length when Stage II crack growth began. The final crack length (a_f) was calculated from the *Damage Tolerant Design Handbook's* K_{IC} value.

The a versus N curves for the various correction factors are shown in Figs. 7 and 8. Each of the curves was started at the initial Stage II crack length or at the lower bound of the correction factor. The Daoud, Carpinteri, and Newman factors all have lower bounds on the correction factor [Newman (a/D) = 0.05, Daoud (a/D) = 0.0625, Carpinteri (a/D) = 0.1]. Both the Murakami and Newman factors have an upper bound on the correction factor [Murakami (a/D) = 0.25, Newman (a/D) = 0.35]. If the polynomial expression of the Murakami correction factor is used past the upper limit, the K predicted by the correction factor begins to drop, invalidating all values after this point. The Newman factor, on the other hand, continues to predict increasing K values past its upper limit and predicts a reasonable final crack size and cycles to failure. However, in keeping with the prescribed bounds, all correction factors are limited to their bounds on both Figs. 7 and 8.

A major shortfall of many correction factors is their limited range of use. Those factors that have an upper bound (Newman and Murakami) are not appropriate when considering fatigue in which there is a significant crack growth. Those models with a lower bound (Newman, Carpinteri, Daoud) cannot be used to predict initiation and total life. In some cases the bounds limit the correction factor to the point that it becomes unfeasible to use it to predict crack growth. Two examples in which the bounds severely limit the correction factor are the Carpinteri and Murkami factors. The lower bound of the Carpinteri correction factor limits its validity to the second half of the fatigue life. The upper bound of the Murkami factor limits its validity to the first half of the fatigue life.

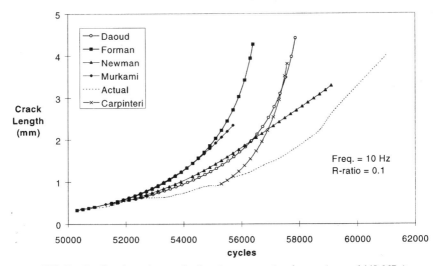

FIG. 7—*Predicted crack growth of various correction factors (σ_B = 1448 MPa).*

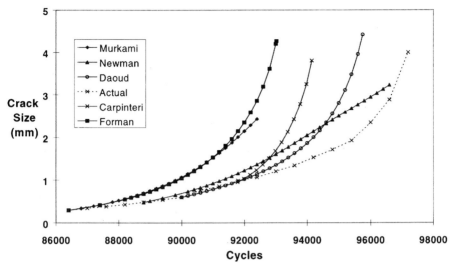

FIG. 8—*Predicted crack growth of various correction factors* ($\sigma_B = 1427\ MPa$).

In general, all the correction factors predict a faster crack growth and a slightly larger final crack size than seen experimentally. The slightly larger crack size predicted by all the correction factors is probably due to the high value of K_{IC} obtained from the *Damage Tolerant Design Handbook*. Table 1 shows the final flaw size and cycles to failure predicted by each of the correction factors. The Forman factor provides the most conservative predictions of crack growth. The Murakami factor predicts similar values as the Forman factor while it is still valid. The Carpinteri factor predicts a very short fatigue life once it valid. The factors that best predict the crack growth are the Newman and Daoud factors. Both of these factors predict a slower rate of crack growth than the others. The Newman factor predicts the longest fatigue life. However, because of an upper bound, it is unable to predict when fracture will occur. Overall, the Daoud factor predicts the second longest fatigue life and it predicts the longest fatigue life of the three factors that are valid when fracture occurs.

There is little change in the correction factors at different loads. The experimental data at 1427 MPa (207 ksi) show a faster rate of crack growth than at 1448 MPa (210 ksi). Since

TABLE 1—*Predicted cycles to failure and final flaw size of each correction factor.*

	Experiment 1 $\Delta\sigma = 210$ ksi		Experiment 2 $\Delta\sigma = 207$ ksi	
	Cycles to Failure	Final Crack Size, mm	Cycles to Failure	Final Crack Size, mm
Daoud	57 872	4.4	95 747	4.4
Newman*	59 100	3.3	96 595	3.2
Forman	56 400	4.2	93 016	4.3
Murakami*	55 705	2.3	92 397	2.4
Carpinteri	57 580	3.8	94 148	3.8
Experimental	59 246	4.0	96 600	4.0

*The correction factor's upper limit was reached before the specimen failed.

all of the correction factors predict this faster rate of crack growth, there is better agreement with the 1427 MPa (207 ksi) data set than with the 1448 MPa (210 ksi) data set.

Fracture Toughness (K$_C$)

The fracture toughness of the bolt could be calculated using

$$K_{IC} = \sigma\sqrt{\pi a_f}F(a) \tag{5}$$

The final crack size was estimated from the fracture surfaces of the bolt. The fracture toughness for three bolts, which had a measurable final crack size, was calculated and compared with the *Damage Tolerant Design Handbook* K_{IC} of 117 MPa $\cdot$ m$^{1/2}$ (106 ksi in.$^{1/2}$). The results are shown in Table 2. The Daoud, Forman, and Carpinteri factors are the only ones that are valid for all three of the final a/D ratios. The Newman correction factor is valid for 1993 MPa (290 ksi) test. The Forman and Daoud equations predict similar results, which are close to the handbook data. Both predict a higher K_C that is worst-case conservative by 20%. The Carpinteri predictions are 20 to 30% lower than the handbook while Newman is 30% lower. If conservative estimates of K_C are desired, then either the Daoud or the Forman factor should be used.

Summary and Conclusion

In summary, two different fatigue life prediction methodologies were compared with literature data. An *S-N* curve for a 4340 steel bolt under bending fatigue was compared with an *S-N* curve for a similar 4340 steel under tensile fatigue from *Mil-Handbook-5*. The bending *S-N* curve showed that bolts have a longer life and higher run-out stress under bending compared with tension.

The second life prediction methodology used was fracture mechanics. Two crack growth tests were run on 4340 bolts under bending. The surface crack length, *c*, was measured and correlated to the inner crack length, *a*, by determining the crack shape throughout the fatigue life of the bolt. The crack shape was found to remain circular until the very last stages of fatigue life. Crack growth data for 4340 plate steel were taken from the *Damage Tolerant Design Handbook* and used to compare with the experimental crack growth data. The handbook data were converted from a *da/dN* versus ΔK curve to an *a* versus *N* curve for each of the correction factors by integrating the Paris equation, $da/dN = C(\Delta K)^m$. All the correction factors conservatively predicted crack growth.

The results lead to the following conclusions:

TABLE 2—*Predicted fracture toughness of 4340 bolts.*

Stress Level, ksi	Final (a/D)	Daoud MPa $\cdot$ m$^{0.5}$	Forman MPa $\cdot$ m$^{0.5}$	Carpinteri MPa $\cdot$ m$^{0.5}$	Newman MPa $\cdot$ m$^{0.5}$
206	0.42	104	108	72	...
208	0.48	123	132	80	...
289	0.32	116	116	89	69

- Tensile S-N data do not accurately predict the bending fatigue life.
- A very conservative estimate of the bending life can be extrapolated from tensile S-N curves.
- Crack growth data from the literature combined with geometric correction factors yield a conservative prediction of crack growth.
- It is unclear which variable, the geometric correction factors or the crack growth data from the *Damage Tolerant Design Handbook,* most influences the conservative predictions.
- Fracture toughness values are very sensitive to the value of the correction factor. Great care should be exercised when choosing a correction factor for this end.
- More work needs to be done on developing correction factors that will better predict crack growth.

References

[1] Ahmad, H., Counts, W., and Johnson, W. S., "Evaluation of Bolt Bearing Behavior of Highly Loaded Composite Joints at Elevated Temperatures," *SAMPE '99,* Society for the Advancement of Material and Process Engineering, Covina, CA.

[2] Si, E., "Stress Intensity Factors for Edge Cracks in Round Bars," *Engineering Fracture Mechanics,* Vol. 37, No. 4, 1990, pp. 805–812.

[3] *Damage Tolerant Design Handbook,* CINDAS/USAF CRDA Handbooks Operation, Purdue University, Vol. 1, May 1994.

[4] *Mil-Handbook-5G,* Metallic Materials and Elements for Aerospace Vehicle Structures, Vol. 1, 1994.

[5] Daoud, O. E. K. and Cartwright, D. J., "Strain Energy Release Rate for a Circular-Arc Edge Crack in a Bar Under Tension or Bending," *Journal of Strain Analysis,* Vol. 20, No. 1, 1985, pp. 53–58.

[6] Forman, R. G. and Shivakumar, V., "Growth Behavior of Surface Cracks in the Circumferential Plane of Solid and Hollow Cylinders," *Fracture Mechanics, ASTM STP 905,* American Society for Testing and Materials, West Conshohocken, PA, 1984, pp. 59–74.

[7] Newman, J. C. and Raju, I. S., "Stress-Intensity Factors for Circumferential Surface Cracks in Pipes and Rods under Tension and Bending Loads," *Fracture Mechanics,* ASTM STP 905, American Society for Testing and Materials, West Conshohocken, PA, 1984, pp. 789–805.

[8] Carpinteri, A., "Elliptical-Arc Surface Cracks in Round Bars," *Fatigue and Fracture of Engineering Materials and Structures,* Vol. 15, No. 11, 1992, pp. 1141–1153.

[9] Murakami, Y. and Tsursu, H., *Stress Intensity Factors Handbook,* Society of Material Science, Japan, 1986, pp. 657–658.

[10] Athanassiadis, A., Boissenot, J. M., Brevet, P., Francois, D., and Raharinaivo, A., "Linear Elastic Fracture Mechanics Computations of Cracked Cylindrical Tensioned Bodies," *International Journal of Fatigue,* Vol. 17, No. 6, December 1981, pp. 553–566.

[11] Levan, A. and Royer, J., "Part-Circular Surface Cracks in Round Bars under Tension, Bending, and Twisting," *International Journal of Fracture,* Vol. 61, 1993, pp. 71–98.

M. Gaudett,[1] *R. Tregoning,*[2] *E. Focht,*[1] *X. J. Zhang,*[1] *and D. Aylor*[3]

Laboratory Techniques for Service History Estimations of High Strength Fastener Failures

REFERENCE: Gaudett, M., Tregoning, R., Focht, E., Zhang, X. J., and Aylor, D., "**Laboratory Techniques for Service History Estimations of High Strength Fastener Failures,**" *Structural Integrity of Fasteners: Second Volume, ASTM STP 1391,* P. M. Toor, Ed., American Society for Testing and Materials, West Conshohocken, PA, 2000, pp. 16–35.

ABSTRACT: Two fastener failures have been analyzed to determine both failure mode and service loading history. The analyses were conducted through careful fractographic investigation and laboratory simulation to assess the integrity of the assembly design and to evaluate the role of unintended environmental effects in these failures. The first case involves Monel K-500 bolts that failed due to fatigue loading, terminated by overload fracture. The fatigue mode was intergranular (IG) at initiation, and then transitioned to classical transgranular fatigue prior to overload failure. Intergranular cracking in Monel K-500 is usually associated with environmentally assisted cracking (EAC). However, simulated service testing of several bolts under both fatigue and slow strain rate loading revealed that IG fatigue cracking could occur at low applied ΔK in air. Transgranular fatigue is associated with higher ΔK levels. Therefore, EAC was not the root cause of these failures.

The second case examines the fatigue failure of several IN625 studs. The stud stresses are estimated from fatigue striation measurements using a representative fatigue crack growth behavior and driving force equation. The inferred applied stress is greater than the material yield strength, which is indicative of a basic deficiency in the original joint design. Laboratory testing verifies the relative accuracy of this stress estimation method for both simple and complex loading histories. The accuracy can be most significantly improved through fatigue crack growth testing of the actual failed material under relevant service conditions.

KEYWORDS: IN625, K-500, high strength fastener materials, failure analysis, environmentally assisted cracking, fatigue

Fastener failure analyses are typically conducted to determine the cause of failure and to determine if the failure can be attributed to one of three possibilities: (1) material deficiency, (2) improper installation, or (3) design inadequacy. Material or installation problems can usually be easily solved by direct replacement. However, failures due to design inadequacy are less easily remedied. An inferior design may require expensive examination and redesign of all similar fastener assemblies. Therefore, it is very important to accurately verify the root cause of failure before determining the proper remedial action.

This paper describes two fastener failures in which design inadequacy was identified as a possible cause of failure. Additional laboratory testing and simulation were performed to not only verify the failure mode, but to infer loading history and assess the integrity of the

[1] Materials engineer, Carderock Division/Naval Surface Warfare Center, Bethesda, MD 20817-5700.
[2] Mechanical engineer, Carderock Division/Naval Surface Warfare Center, Bethesda, MD 20817-5700.
[3] Corrosion engineer, Carderock Division/Naval Surface Warfare Center, Bethesda, MD 20817-5700.

design. Unintended environmental effects and the accuracy of the loading history determination were evaluated by reproduction of the failure mode via laboratory simulation. The failure analyses of Monel K-500 bolts that failed in fatigue are discussed. Intergranular (IG) cracking was identified at the initiation region. Therefore, environmentally assisted cracking (EAC) mechanisms were investigated. Two possible service conditions that may have contributed to intergranular failure were simulated in the laboratory to identify the loading mode that caused failure and the weakness in the assembly design.

Also discussed is the failure analysis of IN625 studs. Once again, fatigue was the primary failure mode. Quantitative fractographic methods were employed to determine the service load magnitude. The accuracy of this technique was verified using both controlled laboratory testing and stud loading simulation.

Intergranular Cracking of Failed Alloy K-500 Fasteners

Background

Monel K-500 fastener[4] failures were observed and brought to the attention of the Naval Surface Warfare Center, Carderock Division (NSWCCD), for failure analysis and remedial recommendations. Wet chemistry analysis showed that the chemical composition of the fasteners met the specification requirements for Alloy K-500.

The fracture surfaces of seven fasteners were examined to determine the failure sequence and initiation mode. Fracture initiated in the third or fourth root from the fastener head in all of the failed fasteners that were examined. The general visual appearance was characteristic of classical fatigue failure. The principal crack plane was flat and perpendicular to the longitudinal axis of the fasteners. In two cases crack initiation sites were noted 180° apart at the circumference and final fracture occurred at the center of the fastener. This indicates that a reversed bending component was applied to these fasteners while in service. In most cases, the principal crack propagated through 60 to 90% of the fastener diameter before deviating out of plane over the remaining ligament where adjacent threads link together.

Observations of the fracture surfaces of the fasteners using the scanning electron microscope (SEM) revealed three distinct fracture modes. The predominant mode was transgranular fatigue crack growth (occupied 80 to 90% of the fracture surface). Microvoid coalescence was evident over the out-of-plane ligament, which is indicative of the final overload fracture region. Intergranular cracking (IG) was also observed on the fracture surfaces adjacent to the thread roots of all seven failed fasteners (Fig. 1). The IG cracking extended from the thread root towards the center for a distance of 100 to 300 μm, but it was not observed to extend completely around the circumference of every fastener. The IG cracking was more prevalent at the apparent crack initiation sites.

Four intact fasteners were examined metallographically to determine if cracking was present only in the failed fasteners. The fasteners were sectioned parallel to the longitudinal axis, then polished and etched to reveal the microstructure. Each fastener exhibited both transgranular and IG cracking emanating from several thread roots. The material microhardness was measured at several thread roots and found to be below 35 HRC.

The cause of the IG cracking was not immediately clear. Alloy K-500 classically exhibits IG cracking due to hydrogen embrittlement (HE) when the material tensile strength is high

---

[4] The available information indicates that fasteners were 0.5-13UNC flat socket head cap screws purchased to MIL-S-1222H.

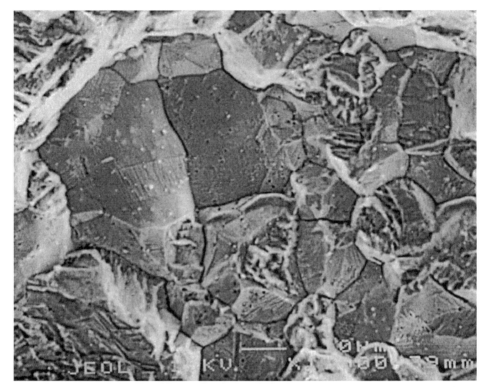

FIG. 1—*Intergranular cracking observed on the fracture surfaces of the failed fasteners. The IG cracking was observed up to 100 to 300 μm from the thread roots.*

[1]. However, the measured microhardness was below the observed HE susceptibility limit (<35 HRC) [1]. Additionally, the cathodic protection (CP) level was reportedly more positive than the observed electrochemical potential threshold below which HE of Alloy K-500 has been observed [2,3]. Experiments were therefore necessary to determine the cause of IG cracking within the expected service environment.

Experimental Design

Two types of tests were performed to isolate the potential influences of the CP level and fatigue loading. Monotonic slow strain rate tests (SSRT) were conducted to examine the effect of environment and CP on the evolution of IG cracking. This allowed the effects of applied stress on hydrogen (H) ingress [4] and local crack tip chemistry to be considered [5] under a slowly varying stress state. Fatigue crack growth experiments were also performed to determine if IG cracking could evolve in the absence of an environmental contribution. Intergranular cracking of K-500 due to fatigue loading has been previously reported in the initial stages of crack growth [6–8].

Pre-Charging Treatments

Reduced section specimens were machined directly from the failed fasteners (Fig. 2) in order to remove potentially embrittled material from the thread root region. The specimens

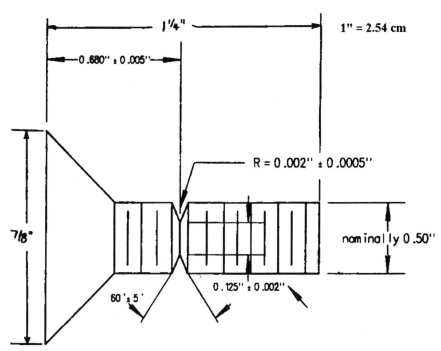

FIG. 2—*Drawing of the specimen used in slow strain rate and fatigue tests. The original fasteners were 0.5 to 13UNC flat socket head cap screws and the cross section was reduced from 1.3 cm to 0.3 cm.*

were pre-charged with H potentiostatically at -800 mV versus a saturated calomel electrode (SCE) in artificial seawater (ASTM Standard Specification for Substitute Ocean Water, D 1141). Hydrogen pre-charging was conducted for pre-stress levels beyond the typical fastener installation level. Table 1 summarizes the charging time, the applied pre-load stresses, and the test performed on each specimen.

Mechanical Testing

Fatigue and SSRT tests were conducted both in laboratory air and in a 3.5% NaCl solution, under CP at -800 mV versus SCE. A potential of -800 mV versus SCE is thought to represent an electrochemical potential "threshold" above which the susceptibility of Alloy K-500 to HE diminishes significantly, and is the reason why this applied level was chosen [9,10]. After specimen failure, one half of each fracture surface was immediately stored in liquid nitrogen for future H level measurements. The hydrogen contents were measured by an outside company using a hot vacuum extraction technique. The SSRT was conducted on two of the four specimens for each pre-charging condition. A constant crosshead displacement rate of 2.3×10^{-5} mm/s was utilized.

Axial fatigue tests were performed on the remaining two specimens for each pre-charging condition. The testing was conducted using a constant R ratio of 0.1, a cyclic frequency of 10 Hz, and various stress amplitudes. The failed fasteners exhibited no signs of net section plastic deformation, indicating that the service loads did not exceed the yield strength (σ_{ys})

TABLE 1—Experimental conditions and a summary of the results of the intergranular cracking investigation of Alloy K-500 fasteners.

		Experimental Conditions				Summary of Results				
Specimen ID	Test	Pre-Charge Time, months	CP Level,[c] mV vs. SCE	Applied Stress During H_2 Charging, MPa	Max. Stress ($R = 0.1$), MPa	NTS, MPa	NTS/ NTSair[a]	Cycles-to- Failure	Hydrogen, ppm	Intergranular Cracking (Extensive, Moderate, Isolated, or None)
NTA	SSRT	3	−0.800	0	N/A	1262	0.90	N/A	2.3	None
NTB	SSRT	3	−0.800	0	N/A	1262	0.90	N/A	...	None
NTC	Fatigue	5	−0.800	0	345	N/A	N/A	647 230	5.4	Moderate/extensive
NTD	Fatigue	5	−0.800	0	345	N/A	N/A	188 670	...	Moderate/extensive
T2A	SSRT	5	−0.800	826.7	N/A	979	0.69	N/A	...	Extensive
T2B	SSRT	5	−0.800	835.0	N/A	1062	0.75	N/A	6.9	Extensive
T2C[b]	Fatigue	5	−0.800	895.7	345	N/A	N/A	...	...	...
T2D	Fatigue	5	−0.800	812.6	345	N/A	N/A	1 020 600	9.2	Isolated
T3A	SSRT	8	−0.800	681.9	N/A	1393	0.99	N/A	7.1	None
T3B	SSRT	8	−0.800	666.1	N/A	1317	0.93	N/A	...	None
T3C	Fatigue	8	−0.800	715.7	345	N/A	N/A	2 849 700	...	Extensive
T3D	Fatigue	8	−0.800	679.2	345	N/A	N/A	261 720	5.9	Extensive
L2A	SSRT	5	−0.800	271.7	N/A	1400	0.99	N/A	...	None
L2B	SSRT	5	−0.800	268.9	N/A	1400	0.99	N/A	3.3	None
L2C	Fatigue	5	−0.800	268.9	345	N/A	N/A	126 760	...	Moderate
L2D	Fatigue	5	−0.800	270.3	345	N/A	N/A	230 020	3.2	Moderate
L2E	SSRT	8	−0.800	274.4	N/A	1386	0.98	N/A	...	Isolated

Specimen	Test		CP[c]							
L2F	SSRT	8	−0.800	275.1	N/A	1351	0.96	N/A	5.5	Isolated
L2G	Fatigue	8	−0.800	277.2	345	N/A	N/A	211 560	7.5	Moderate
L2H	Fatigue	8	−0.800	268.9	345	N/A	N/A	183 190	...	Moderate
CPA	SSRT	0	−0.800	N/A	N/A	1476	1.05	N/A	0.8	None
CPB	SSRT	0	−0.800	N/A	N/A	1448	1.03	N/A	...	None
CPC	Fatigue	0	−0.800	N/A	345	N/A	N/A	564 870	...	Extensive
CPD	Fatigue	0	−0.800	N/A	345	N/A	N/A	551 290	1.1	Extensive
LAA	SSRT	0	Air	N/A	N/A	1467	1.04	N/A	...	None
LAB	SSRT	0	Air	N/A	N/A	1351	0.96	N/A	...	None
LAC	Fatigue	0	Air	N/A	552	N/A	N/A	36 023	0.5	None
LAD	Fatigue	0	Air	N/A	552	N/A	N/A	21 705	0.7	None
LAE	Fatigue	0	Air	N/A	345	N/A	N/A	5 992 600	...	Run out
LAF	Fatigue	0	Air	N/A	345	N/A	N/A	2 310 400	...	Isolated
LAG	Fatigue	0	Air	N/A	655	N/A	N/A	15 769	...	Isolated
LAH	Fatigue	0	Air	N/A	655	N/A	N/A	17 294	...	Isolated

[a] NTSair is the average NTS for specimens LAA and LAB.
[b] Specimen T2C was broken during installation into the test frame.
[c] The CP level was maintained during the 3, 5, or 8 month pre-charging and subsequent SSRT and fatigue testing.

of the material (σ_y = 690 MPa). Therefore, maximum stresses of 345, 552, and 655 MPa were arbitrarily chosen for the fatigue testing (Table 1).

Results and Discussion

H Measurements—The results are shown in Table 1 and Fig. 3. The highest bulk level of H was measured in the specimens with the highest pre-load during charging (T2B and T2D). The lowest measured H levels were obtained from the specimens that were never exposed to chloride conditions or only exposed during testing. Figure 3 shows that the level of H increased with increasing exposure time, as expected. Pre-load stresses less than 679 MPa do not increase the H ingress rate. However, when the pre-load stress is higher than the yield strength (690 MPa), the H content is significantly greater than the upper 95% confidence interval calculated for the remaining data (Fig. 3).

Slow Strain Rate Tensile Testing—Table 1 contains a summary of the slow strain rate test results performed on the fastener specimens. Specimens that were either pre-charged but not pre-stressed (NTA and NTB), or pre-stressed up to 682 MPa (L2A, L2B, L2E, L2F, T3A, and T3B) did not show significant IG cracking. Only isolated patches of intergranular cracking were observed on specimens L2E and L2F after a pre-charge of eight months. Specimens T2A and T2B were pre-charged under high stress conditions (827 MPa) for five months, at stress levels above the yield strength of Alloy K-500, and are the only SSRT specimens to exhibit extensive intergranular cracking (Fig. 4).

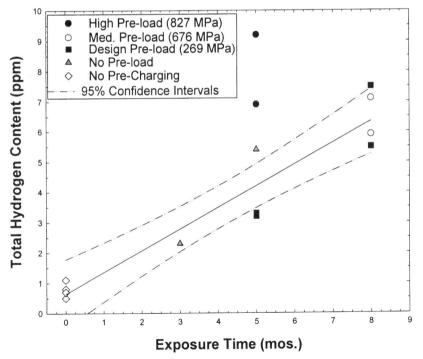

Exposure Time (mos.)

FIG. 3—*Total hydrogen content as a function of pre-charging exposure time measured from the fractured SSRT and fatigue specimens.*

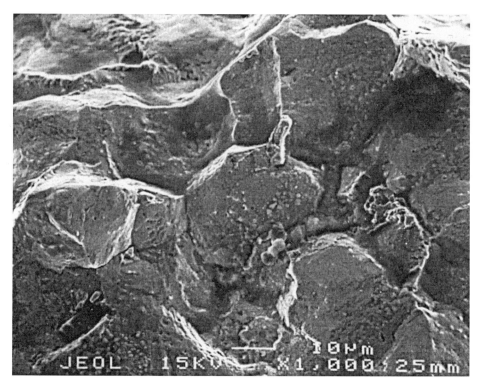

FIG. 4—*SEM fractographs of specimen T2A, pre-loaded to 827 MPa, and pre-charged for five months. Extensive IG cracking was observed near the root of the notch.*

The results of the slow strain rate tensile testing performed in this study showed that the pre-charging duration did not appear to have a controlling effect on the amount of IG cracking. In contrast, H measurements and the SSRT tests show that the level of the pre-stress appeared to increase the level of H uptake (Fig. 3) and the amount of IG cracking, but only when the pre-stress exceeded yield.

Axial Fatigue Testing—Table 1 shows the results of the fatigue testing performed on the K-500 specimens. The results in Table 1 show that IG cracking was observed both on specimens that were tested in air (Fig. 5) and in a 3.5% NaCl solution at −800 mV versus SCE (Fig. 6). The amount of IG cracking appeared to increase when the specimens were tested in the 3.5% NaCl solution under CP. In addition, Table 1 indicates that the pre-charging duration did not have a significant effect on the amount of IG cracking that occurred during fatigue testing. This conclusion is further support by the extensive IG cracking observed in specimens that were not pre-charged and tested in the 3.5% NaCl solution at −800 mV versus SCE (CPC and CPD).

The most significant result of these experiments was the excessive amount of IG cracking observed in the fatigue specimens compared to the SSRT specimens, with the exception of SSRT specimens T2A and T2B mentioned above. It was shown here that IG cracking could be obtained during fatigue crack growth for a stress magnitude that is well below yield, while yielding is required to generate IG cracking in SSRT. In addition, the lack of correlation between measured average H contents and the amount of IG cracking observed in the fatigue

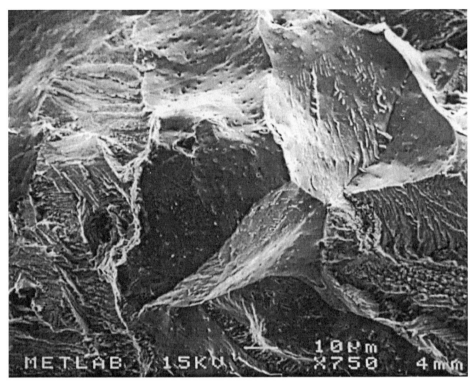

FIG. 5—*SEM fractograph of fatigue specimen LAF tested in air with no pre-load or pre-charging. Isolated IG cracking was observed.*

tests does not support a bulk HE fracture mechanism for these fasteners. Instead, it appears that H-assisted cracking is controlled by the local H content at the crack tip.

It has been reported that bulk HE may not be necessary to cause IG cracking in Alloy K-500 [11]. This work has suggested that the HE observed in their experiments was a crack-tip mechanism rather than a bulk HE process. Such a scenario may help to explain IG cracking observed at lower stress intensities. At the slow crack growth rates associated with low stress intensities, there is more time for H accumulation in the fracture process zone (FPZ). In addition, the FPZ is small and plastic deformation can interact locally with grain boundaries to contribute to IG fracture. As the crack growth rate increases with stress intensity, there is insufficient time available to accumulate a critical H content in the FPZ. In addition, the FPZ is large and plastic deformation interacts globally with the microstructure. Therefore, a transition from IG to transgranular fracture occurs with increasing ΔK. The experiments described here were not carried out to confirm such a scenario. However, tests are currently underway to determine the conditions that lead to IG cracking during fatigue in terms of environment and the applied stress intensity factor range (ΔK).

Results published in the open literature [6–12] indicate a propensity for pure Ni, K-500, and other nickel-based alloys to exhibit IG cracking during fatigue. The results indicated that IG cracking was most likely to occur during the initial stages of crack growth when the ΔK was low and transition to transgranular cracking as the crack length and ΔK increased. It was postulated that strain concentration at the grain boundaries results from the formation

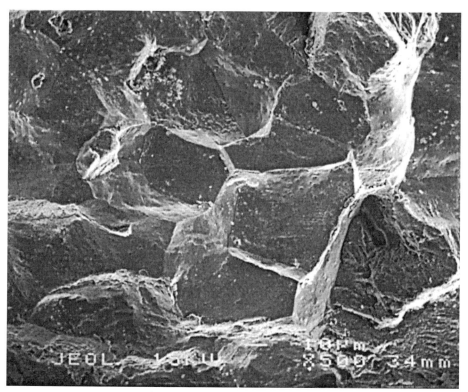

FIG. 6—*SEM fractograph of fatigue specimen L2D, pre-loaded to 269 MPa, and pre-charged for five months. A moderate level of IG cracking was observed.*

of persistent slip bands in the locally softer area near grain boundaries [7,12]. It may be possible that the uptake of H exacerbates this process by mechanisms such as slip localization [13] and grain boundary pileup stress enhancement [14], softening of the grain boundary vicinity [15], or the reduction of grain boundary cohesion [16]. More work needs to be performed to identify the H-assisted fatigue fracture mechanism in Alloy K-500.

Summary

Fatigue crack growth and SSRT were used to identify the cause of IG cracking in several failed Monel K-500 bolts. The imposed pre-charging conditions caused extensive HE during SSRT only when the pre-load stress exceeded the material yield strength. This loading magnitude is extreme compared to typical service conditions. The pre-charging duration did not have a significant effect on the level of IG cracking in either the SSRT or fatigue tests.

Isolated IG fatigue cracks were obtained in notched Alloy K-500 specimens tested in air under axial fatigue at a maximum stress of about 345 MPa ($R = 0.1$). The simulated service environment increased the susceptibility of Alloy K-500 to IG cracking under fatigue loading, but was not necessary to develop this failure mode. For the fatigue specimens that exhibited moderate to extensive amounts of IG cracking, it is inferred that the fatigue crack propagation mode was IG at low ΔK levels, and transgranular at high ΔK.

Considering the body of evidence, both from the work performed herein and from the literature, the IG cracking observed on the fracture surfaces of the failed K-500 fasteners was more likely due to fatigue and/or corrosion fatigue and less likely due to bulk HE. However, HE should always be considered in cases where Alloy K-500 fasteners are being used in applications requiring CP.

Failure Analysis of a Fractured Inconel 625 Stud

Optical and SEM fractographic observations were performed on the fracture surface of a failed Inconel 625 (IN625) stud. The nominal diameter of the stud was 22 mm with 3.5 threads/cm (9 threads/in.). It is unknown whether the threads were cut or rolled. Fatigue cracking initiated at numerous sites near the thread root and propagated over a substantial portion of the stud cross section (65% in area). Closer examination revealed that the fatigue region actually consisted of several alternating bands of fatigue and ductile fracture. These regions were separated by bench marks and indicate changes in the applied loading. The fatigue cracking was terminated by tensile overload fracture on the remaining cross section ligament. No evidence of environmentally assisted fatigue or fracture was observed.

This apparently high-stress failure mechanism raised concerns about the adequacy of the structural design, as the expected stresses were low. It was therefore decided to estimate a simplified loading history using quantitative fractography. The accuracy of these stresses was evaluated using laboratory simulation. Several threaded rod specimens were tested at NSWCCD under known, axial stress ranges at a constant $\sigma_{min}/\sigma_{max}$ (R ratio) to provide a simple comparison with stresses estimated from the failure surface. Additionally, independent testing under simulated service conditions was conducted under undisclosed loading conditions. The stresses from these tests were also estimated using quantitative fractography. The actual loading history for the "blind testing" was provided after the fractographic stress estimation was completed. The results show that analytical fractography is accurate for predicting relatively simple loading scenarios and is qualitatively useful for bounding and ranking more complicated service loading conditions.

Stress Determination Using Quantitative Fractography

General Approach—The stress range is determined using fatigue striation and crack length measurements on the fracture surface in combination with a published crack growth law and an idealized driving force relationship. Fatigue striation measurement is conducted to determine the crack growth per cycle (da/dN) at various locations on the failed component. Additionally, the crack depth (a) and crack surface length (b) are also measured at each site. A relevant fatigue crack growth rate (FCGR) relationship is then chosen or measured for the material studied. The stress intensity factor range (ΔK) is estimated from this relationship at the measured da/dN values. The stress range ($\Delta\sigma$) can then be determined using a generalized driving force equation of the form

$$\Delta\sigma = \frac{\Delta K}{Y(a,b,D)\sqrt{\pi a}} \tag{1}$$

where Y is the shape factor for a given geometry (i.e., minimum bar diameter D) and loading condition [17,18]. The $\Delta\sigma$ value is determined at each fatigue striation site.

If it is known that the applied stress range is constant during service, $\Delta\sigma$ at each measurement point can be averaged and utilized as a nominal value. Additionally, the maximum applied stress magnitude can be estimated from the stress at ductile overload (σ_f). This stress is approximated using the true failure tensile stress (σ_{brk}) from a tensile test, and the ratio of final thread root ductile fracture area to shank area (A_f/A_o). Then, if σ_{brk} is equivalent for the stud and tensile test

$$\sigma_f = \frac{A_f}{A_o} \sigma_{brk} \qquad (2)$$

The maximum applied stress (σ_{max}) is therefore equal to σ_f, while the minimum applied stress (σ_{min}) is equal to ($\sigma_f - \Delta\sigma$).

This methodology is presented in detail for the failed IN625 stud. This methodology does assume a loading history during which the applied stress range is constant. As mentioned previously, the appearance of ductile fracture indicates periodic load increases during which the final fracture of the stud likely took place, indicating that a constant stress range was *not* obtained. However, this technique is employed here to obtain an estimation of the upper bound of the stress range seen by the stud.

Application to IN625 Stud Failure—The fatigue striation spacing was measured at five locations near the thread roots. The locations were near the initiation region between $0.02 < a/D < 0.15$. The average spacing is 0.1 μm at all locations (Fig. 7) which implies that $da/dN \approx 0.1$ μm/cycle.

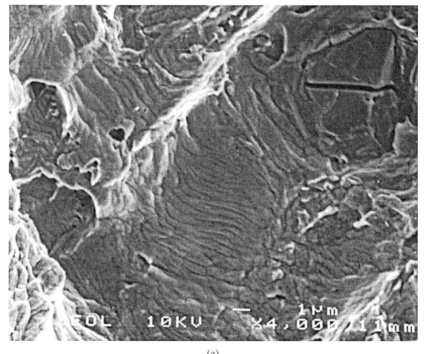

(a)

FIG. 7—*Fatigue striations on an IN625 stud fracture surface at* a = *1.75 mm:* (a) *low magnification, and* (b) *high magnification.*

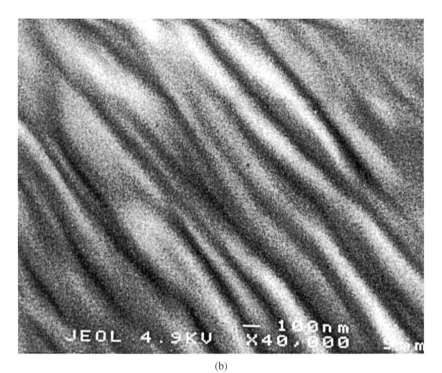

(b)

FIG. 7—Continued.

A FCGR curve for compact tension tests conducted in 3.5% NaCl solution, at a constant 10 Hz test frequency, under constant R ($\sigma_{max}/\sigma_{min}$) = 0.1 ratio loading (uncorrected for closure effects) was obtained from the literature for IN625 plate with the following mechanical properties: a yield strength of 486 MPa, tensile strength of 949 MPa, and an elongation at failure of 43% (19,20). The FCGR data was used to estimate the applied ΔK. For the measured da/dN = 0.1 μm/cycle, ΔK was shown equal to approximately 24 MPa$\sqrt{m}$.

The following stress intensity factor solution for a straight crack in a threaded round bar under tensile loading was utilized [17,18]

$$\frac{\Delta K}{\Delta\sigma\sqrt{\pi a}} = 2.043 \exp\left[-31.332\left(\frac{a}{D}\right)\right] + 0.6507 + 0.5367\left(\frac{a}{D}\right)$$

$$+ 3.0469\left(\frac{a}{D}\right)^2 - 19.504\left(\frac{a}{D}\right)^3 + 45.647\left(\frac{a}{D}\right)^4 \quad (3)$$

This empirical solution was formulated from several different finite element and experimental studies [17]. This solution includes stress concentration effects from the thread roots and is reported to be valid for $a/D > 0.004$. The pure cyclic tensile loading approximation was chosen because of the equiaxed appearance of the microvoids observed on the failed stud. The stress range calculated using these described assumptions and inputs is approximately 280 MPa.

The final applied remote stress (σ_f) was calculated for this material using a conservative estimation of σ_{brk} (the true failure stress for IN625). The true failure stress (σ_{brk}) can be calculated using the following equation [21]

$$\sigma_{brk} = S_f(1 + e_f) \qquad (4)$$

where S_f is the engineering stress at failure, and e_f is the elongation at failure. A conservative estimate for S_f is obtained by using a typical tensile strength (965 MPa) and e_f(0.45) for IN625 [19,20,22]. Using Eq 4, a conservative estimate of 1380 MPa is obtained for σ_{brk}, and is used throughout this paper to determine σ_f.

The A_f/A_o ratio as measured by low power microscopy was approximately 0.35 for the failed stud. Therefore, using Eq 2, $\sigma_f = \sigma_{max} = 480$ MPa. The mean stress (σ_{mean}) calculated from σ_{max}(480 MPa) and $\Delta\sigma$ (280 MPa) is 340 MPa. This value is consistent with a typical pre-load stress of 2/3 of the material yield strength, equal to 2/3 (486 MPa) = 324 MPa.

Experimental Evaluation of Fractographic Stress Estimation

NSWCCD Verification Testing—Three uncracked stud sections were obtained from the fractured IN625 studs. The specimens were located away from the original fracture. The nominal diameter was identical to the failed studs (22 mm) with 3.5 threads per cm. The fatigue specimens were tested in tension at three different constant stress ranges. The stress ranges were chosen to simulate the large applied maximum and alternating stresses predicted for the failed studs. Once again, quantitative fractography was used to estimate σ_{max} and $\Delta\sigma$ independently of the known test conditions.

Table 2 shows the cyclic test conditions for the three samples. Specimen N8 was tested with σ_{max} nearly equal to $1.4\sigma_{ys}$. This specimen exhibited a stable fatigue crack growth stage and had a short fatigue life. Specimen N1 was tested at an intermediate stress magnitude and range with $\sigma_{max} = 0.85\sigma_{ys}$ and exhibited an intermediate life. Specimen N15 was tested with $\sigma_{max} = 0.60\sigma_{ys}$ (520 MPa) and its fatigue life was the highest.

Striation spacing measurements were made at 5 to 6 crack depths (a) for each sample. The measurements were spaced throughout the stud cross section, but at locations away from the thread root to diminish the effects of any stress concentration. The stress range was calculated at each location following the general approach outlined earlier. The stresses were then averaged to determine a single estimated stress range for each specimen. The maximum stress at ductile overload (σ_f) was once again estimated using the measured A_f/A_o ratio with an assumed σ_{brk} of 1380 MPa.

The results of the stress estimates from quantitative fractography are summarized in Table 3. The actual $\Delta\sigma$ is higher than the estimated $\Delta\sigma$ in every case. However, it is interesting to note that the estimated striation measurements for $a/D < 0.2$ are generally the closest to

TABLE 2—*Test matrix for three NSWCCD IN625 studs with a nominal diameter of 22 mm and 3.5 threads per cm.*

Specimen	σ_{max}, MPa	$\Delta\sigma$, MPa	R	Frequency, Hz	Cycles to Failure, N
N8	720	645	0.1	10	5 600
N1	440	397	0.1	10	140 300
N15	320	289	0.1	20	7 648 200

TABLE 3—*Quantitative fractographic stress estimates for three IN625 studs tested at NSWCCD with a nominal diameter of 22 mm and 3.5 threads per cm* (D = *minimum diameter*).

Specimen	a, mm	a/D	$da\,dN$, μm/cycle	$\Delta\sigma$, MPa	Ave $\Delta\sigma$, Actual (MPa)	A_f/A_o, %	σ_{max}, Actual (MPa)
N8 D = 18.3 mm	1.9	0.11	0.12	455	521 (645)	66	910 (720)
	2.0	0.11	0.25	552			
	2.2	0.12	0.31	552			
	2.9	0.16	0.59	600			
	3.9	0.22	1.40	510			
	4.9	0.28	2.11	455			
N1 D = 17.9 mm	2.5	0.14	0.10	359	280 (397)	42	579 (440)
	2.8	0.15	0.13	365			
	3.5	0.19	0.17	310			
	5.5	0.30	0.45	228			
	7.5	0.41	0.92	241			
	9.2	0.50	1.60	179			
N15 D = 20.0 mm	3.2	0.16	0.10	310	200 (289)	28	386 (320)
	4.2	0.21	0.11	296			
	6.2	0.31	0.11	159			
	8.2	0.41	0.25	145			
	10.2	0.51	0.36	110			

the actual values. The estimated values for $a/D > 0.2$ are much lower than the actual values. The predicted stress decreases as a/D and ΔK increases in these constant stress range tests. This implies that either the assumed FCGR behavior for this material is not representative of the actual material behavior at high ΔK, or that the driving force equation becomes less accurate as a/D increases. Most importantly, these validation tests imply that the actual $\Delta\sigma$ value applied to the studs that failed in service is likely higher than the estimated value of 280 MPa.

Conversely, due to differences in the assumed and actual values for σ_{brk}, the estimated σ_{max} values are higher than the actual values. Since the applied σ_{max} is known for each test, the material σ_{brk} values can be determined using Eq 2, and are equal to 1090 MPa (N8), 1050 MPa (N1), and 1035 MPa (N15) for each test. These values are consistent, but less than the assumed value estimated from nominal tensile test results. The lower breaking stress in the threaded fatigue specimen is expected since the higher specimen constraint tends to decrease the ductility at failure. However, the consistency in the σ_{brk} determined for the various specimens implies that this method can accurately determine service σ_{max} values if the true σ_{brk} can be determined or estimated for the service conditions. Alternatively, σ_{max} values based on the tensile σ_{brk} should be conservative as long as component constraint is greater than in a uniaxial tensile test.

Independent Testing—Stress Estimation—An independent party tested five IN625 studs identical to the studs that failed in service (a nominal diameter of 22 mm with 3.5 threads per cm). These studs were tested under fatigue conditions until final fracture. The fractured studs were evaluated by NSWCCD to estimate the applied stress magnitude and range. However, no test details were initially provided. Stress estimations were once again performed using the same fractographic method applied to the failure analysis and NSWCCD verification testing. Fatigue striation widths were measured at four identical crack depths for each specimen, and the applied stress range was determined at each position. The results are summarized in Table 4.

Generally, the striation width ranged from 0.1 μm to a few microns. The striation width initially increased with distance from the initiation at the thread root, as would be expected under constant stress loading. However, the width increase during the last several measurements (a = 3 to 9.2 mm) was not substantial enough to imply constant stress loading. In fact, the measurements indicate that stress range decreased dramatically as the cracks approached final depths just before failure (Table 4 and Fig. 8). This observation is supported by the predicted σ_{max} at failure, which is much lower than the predicted applied stress ranges early in the crack growth history.

Also evident in these results is the difference between the predicted loading history among the five specimens. This is obvious in the plot of the estimated stress range ($\Delta\sigma$) as a function of a/D shown in Fig. 8. Because $\Delta\sigma$ appeared to change dramatically, it is not possible to determine the absolute σ_{max} and σ_{min} values during the initial test portion.

Independent Testing Loading Conditions—After the fractographic applied stress estimates were complete, the testing conditions and fatigue life results of the five independently tested fatigue specimens were revealed. The independent test setup was intended to simulate the actual service condition of a stud [23]. A stud was pre-loaded to a constant installation torque by applying a force against a soft bushing. Alternating cantilever loading with R = 0 was applied at the end of a stud in a direction perpendicular to its longitudinal axis (Fig. 9). Testing was conducted on several of these assemblies in parallel using a pressurized fluid

TABLE 4—*Fractographic stress estimation for independent testing of IN625 studs with a nominal diameter of 22 mm and 3.5 threads per cm.*

Specimen ID	Crack Depth, a, mm	da/dN, μm/cycle	ΔK MPa$\sqrt{m}$	$\Delta\sigma$, MPa	A_f/A_o, %	σ_{max}, MPa
EA	0.3	0.12	29	510		
	1	1.3	58	945	31	430
	3	2	65	889		
	9.2	2.2	70	234		
E1	0.3	0.1	24	434		
	1	0.22	34	552	25	340
	3	0.47	43	572		
	9.2	1.08	56	186		
E15	0.3	0.12	27	386		
	1	0.16	29	538	11	150
	3	0.39	40	552		
	9.2	0.49	42	138		
E8	0.3	0.09	24	441		
	1	0.12	27	517	24	330
	3	0.18	30	455		
	9.2	0.47	43	138		
E7	0.3	0.07	23	393		
	1	0.09	24	386	10	140
	3	0.09	24	359		
	9.2	0.1	25	83		

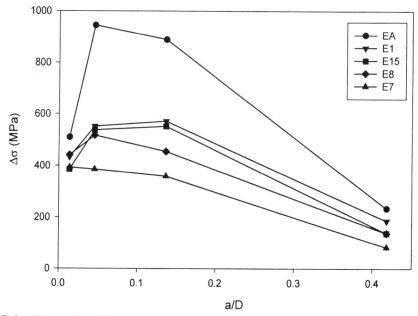

FIG. 8—*Estimated applied stress range* ($\Delta\sigma$) *as a function of crack length* (a) *normalized by minimum diameter* (D) *for IN625 studs tested by an independent party.*

applied by a single pump through a manifold. The maximum applied pump pressure was fixed during the test. This was converted to an applied load (P_{max}) based on the piston area at the assembly load point.

The maximum test load and fatigue life for each sample is summarized in Table 5. The specimens have been listed in order of increasing fatigue life. Also listed are the estimated final σ_{max} and the average $\Delta\sigma$ from the three striation measurements closest to initiation (Table 5). These three measurements were chosen because nominal stress range is nearly consistent for each test.

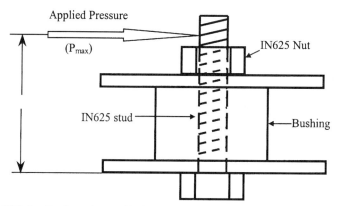

FIG. 9—*Single stud assembly for simulation testing by independent party.*

TABLE 5—*Comparison of reported test conditions and fractographic estimates for IN625 studs with a nominal diameter of 22 mm and 3.5 threads per cm.*

Specimen ID	Fatigue Life, cycles	P_{max}, kN	Average $\Delta\sigma$, MPa	σ_{max}, MPa
EA	19 559	33.4	780	430
E1	81 135	14.2	520	340
E8	249 443	22.2	470	330
E15	604 853	8.9	490	150
E7	917 740	14.2	380	140

It is interesting to note there is a poor general correlation between the applied P_{max} and fatigue life. While variability in the incubation cycles required prior to crack initiation is expected, the difference in fatigue life between specimens E1 and E7, which had similar P_{max} values, appears to be extreme. Both the average $\Delta\sigma$ and σ_{max} predictions do correlate well with the ranking expected based on the actual fatigue life. Specimens with higher average $\Delta\sigma$ and σ_{max} have lower fatigue lives as should be expected. Specimen E8 does have a slightly lower $\Delta\sigma$ than specimen E15, and the final σ_{max} is twice as high. It is likely that this specimen had a higher mean load that compensated for a slightly lower $\Delta\sigma$.

The poor P_{max} correlation is likely due to inconsistent applied loading conditions during the test. First, the actual applied pre-load stress was not measured for each stud. Also, it is unknown how consistent the mean stress was among the samples or if pre-load relaxation due to the soft bushing occurred during the test. Additionally, only the applied pump pressure remained constant during the test, and load was distributed in parallel to several test machines. Specimen cracking or other compliance variability could lead to load redistribution among all the test specimens. Therefore, it is likely that the actual loading history is complicated for this simulation testing, and P_{max} determined from pump pressure alone does not indicate the complicated load history. Though complicated, the applied stress history is consistent with the earlier $\Delta\sigma$ predictions.

Summary

This study shows that the fatigue stress range and maximum stress can be estimated by quantifying the fracture surface features and the use of representative handbook properties. Quantitative fractography indicates that maximum service load is well above the material yield strength and the applied alternating stress range is unacceptably large for the failed IN625 stud. Verification testing illustrates that this method is reasonably accurate for simple loading schemes and can be used to identify and rank failures due to complex loading histories. For the independent simulation testing summarized in Table 4, the method more accurately correlates with fatigue lives than with the reported remote loads. However, only the stress ranges are unambiguously determined from striation width measurements under complex loading. Additional information, such as the overload fracture area, is required to quantify the maximum or mean load.

Uncertainties exist in the method due to variability in the local striation width measurements, available fatigue crack growth rate relationships, and inaccuracies in the estimated driving force equations. Striation width accuracy is improved by conducting measurements at many similar crack depth locations and determining the average growth rate. Rigorous estimates of the driving force relationship will also improve the estimated stress accuracy. However, the largest uncertainty results from the fatigue crack growth rate relationship, which should be determined from tests on the actual material studied. Experiments should be conducted within the expected service environment, loading rate, and R ratio, if these variables are known.

Conclusions

1. The two fastener failures described herein demonstrate the use of laboratory simulation and fractographic analysis to estimate the loading conditions of the fasteners prior to failure. The initial failure analyses suggest deficiencies in the assembly designs.
2. SSRT testing of Alloy K-500 fastener material indicated that an environmental effect could be observed only when the specimen was charged at a stress exceeding the yield stress, which is an extreme condition compared with service conditions.
3. Fatigue testing of Alloy K-500 fastener material indicates that the intergranular fracture observed on the fracture surfaces of the failed K-500 fasteners was more likely due to fatigue and/or corrosion fatigue, and not likely due to an environmental effect alone.
4. Fatigue striation measurements and estimations of the applied stress range for failed studs of IN625 indicate that the fatigue stresses were likely too high for the joint design.
5. The fractographic technique employed to estimate the loading history of the IN625 studs was validated using both NSWCCD and independent laboratory testing, and was determined to be conservative. The usefulness of this technique depends on the validity of the assumptions made in the laboratory tests and analyses, as well as on the relevancy of the available material property information.

References

[1] Efird, K., *Materials Performance,* April 1985, pp. 37–40.
[2] Tregoning, R., Aylor, D., and Caplan, I., "High Strength Alloys for Navy Fastener Applications," *8th World Conference on Titanium Proceedings,* Oct. 1995, Birmingham, England.
[3] Harris, J., Scarberry, R., and Stephens, C., *Corrosion,* Vol. 28, 1972, pp. 57–62.
[4] Chopra, R. and Li, J., Vol. 6, 1973, *Scripta Metallurgica,* pp. 543–545.
[5] Lillard, J., Kelly, R., and Gangloff, R., "Effect of Electrode Potential on Stress Corrosion Cracking and Crack Chemistry of a Nickel-Base Superalloy," *Corrosion 97,* New Orleans, LA, March 1997.
[6] Price, C., *Fatigue and Fatigue and Fracture of Engineering Materials and Structures,* Vol. 11, 1988, pp. 483–491.
[7] Price, C. and Henderson, G., *Fatigue and Fracture of Engineering Materials and Structures,* Vol. 11, 1988, pp. 493–500.
[8] Price, C. and Houghton, N., *Fatigue and Fracture of Engineering Materials and Structures,* Vol. 11, 1988, pp. 501–508.
[9] Joosten, M. and Wolfe, L., "Failures of Ni-Cu Bolts in Subsea Applications," *Proceedings of the 19th Annual Offshore Technology Conference,* Houston, TX, 1987, pp. 29–34.
[10] Aylor, D., "High Strength Alloys for Navy Fastener Applications," CARDIVNSWC-TR-61-95/03, Naval Surface Warfare Center, Carderock Division, West Bethesda, MD, 1995.
[11] Vassilaros, M., Juers, R., Natishan, M., and Vasudevan, A., "Environmental Slow Strain Rate J-Integral Testing of Ni-Cu Alloy K-500," *ASTM STP 210,* R. Kane, Ed., American Society for Testing and Materials, West Conshohocken, PA, 1993, pp. 123–133.
[12] Hwang, K., Plichta, M., and Lee, J., *Materials Science and Engineering,* Vol. A114, 1989, pp. 61–72.
[13] Birnbaum, H. and Sofronis, P., *Materials Science and Engineering,* Vol. A176, 1994, pp. 191–202.
[14] Stroh, A., *Proceedings of the Royal Society of London,* Vol. 223, 1954, pp. 404–414.
[15] Tabata, T. and Birnbaum, H., *Scripta Metallurgica,* Vol. 18, 1984, pp. 231–236.
[16] Bruemmer, S., Jones, R., Thomas, M., and Baer, D., *Metallurgical Transactions A,* Vol. 14A, 1983, pp. 223–232.

[*17*] James, L. and Mills, W., "Review and Synthesis of Stress Intensity Factor Solutions Applicable to Cracks in Bolts," HEDL-SA-3656FP, U.S. Department of Energy, 1987.
[*18*] Popov, A. and Ovchinnikov, A., *Strength of Materials,* Vol. 15, 1983, pp. 1586–1589.
[*19*] Long, T., "A Comparison of Fatigue Crack Propagation in Inconel 625 and 3.25 Ni Steel," Masters thesis, Massachusetts Institute of Technology, March 1977.
[*20*] James, L., "The Effect of Temperature Upon the Fatigue Crack Propagation Behavior of Inconel 625," Report HEDL-TME,77-2, Westinghouse Hanford Co., Richland, WA, 1977.
[*21*] Courtney, T., *Mechanical Behavior of Materials,* McGraw-Hill Publishing Company, New York, 1990, p. 10.
[*22*] *Alloy Digest,* "Inconel Alloy 625," Filing Code Ni-121, Engineering Alloys Digest Inc., Upper Montclair, NJ, 1967.
[*23*] Liu, A., "Behavior of Fracture Crack in a Tension Bolt," *Symposium on Structural Integrity of Fasteners, ASTM STP 1236,* American Society for Testing and Materials, West Conshohocken, PA, 1995, pp. 126–138.

Christopher Wilson[1] and Stephen Canfield[1]

Assembly Cracks in a Hybrid Nylon and Steel Planter Wheel

REFERENCE: Wilson, C. and Canfield, S., **"Assembly Cracks in a Hybrid Nylon and Steel Planter Wheel,"** *Structural Integrity of Fasteners: Second Volume, ASTM STP 1391,* P. M. Toor, Ed., American Society for Testing and Materials, West Conshohocken, PA, 2000, pp. 36–47.

ABSTRACT: In a typical bolted joint, cracks can exist in the fastener, the nut, and either of the two members included in the grip of the fastener. In this paper, the assembly of a hybrid nylon and steel agricultural wheel is studied. Three different cracking problems in the nylon wheel half occurred in the current assembly process. A case study is presented in which the cracking problems arising during the manufacturing and assembly of the hybrid nylon and steel agricultural wheels are addressed. Initially, the cause was thought to be a material processing or fracture mechanics problems. A combination of testing and analysis showed that the root cause was inadequate machine design. After a redesign, cracking problems were eliminated.

KEYWORDS: nylon, assembly cracks, design of agricultural wheels, bolted joints

Gage wheels are used in agricultural planters to cover seed with a precise amount of soil. Conventional gage wheels are made with deep-face (concave) steel surfaces that collect soil thrown on opposing seed rows. The soil builds up and eventually drops to the ground causing an imprecise amount of seed coverage. This additional soil leads to delayed plant penetration and reduced crop yield. A combination nylon and steel gage wheel, shown without the tread in Fig. 1, reduces soil buildup [*1*]. Consistent use of such wheels reduces the variability in plant emergence.

The combination nylon and steel wheel consists of a molded nylon wheel half bolted to a stamped steel half. The thickness of the nylon varies from 0.1 in. (0.254 cm) for the reinforcing ribs to 0.2 in. (0.508 cm) for the bolt seat. The thickness of the steel is 0.1 in. The gage wheel is 16 in. in diameter and has a rubber tread of 4.5 in. The wheel, shown in Fig. 2, is assembled using 0.375-in.-diameter steel bolts in two bolt circles: an outer circle of diameter 10.5 in. with eight equally spaced bolts and an inner circle of diameter 3 in. with three equally spaced bolts. Washer-face locknuts were used to ensure the integrity of the bolted connections. Either cadmium or zinc coated bolts were used. A wheel bearing is seated between the nylon and steel halves within the inner bolt circle.

The original assembly process consists of two stages. In the first stage, a wheel bearing was pressfit into the nylon wheel half. The nylon wheel half was then pressfit into the rubber tread. In the second stage, a set of eleven nuts was placed in a circular jig. The steel wheel half was placed on the jig. Next, the nylon wheel half was placed on its mating steel wheel half. Three alignment pins were seated in the inner bolt circle holes. The nylon wheel and

[1] Assistant professors, Department of Mechanical Engineering, Tennessee Technological University, Cookeville, TN 38505-0001.

FIG. 1—*Nylon and steel wheel without tread.*

rubber tread subassembly was then pressfit onto the steel wheel half. The bolts were dropped into place and a pneumatic torque gun was used to tighten the bolt-heads to an initial torque level of 200 in.-lb. The inner bolt circle was tightened first. Within twenty-four hours, the torque measured on the bolt-heads relaxed to an average value of 140-in.-lb and a minimum value of 120 in.-lb. The 120 in.-lb torque level was considered to be the delivery torque. The 200 and 120 in.-lb torque levels were developed by a combination of experience with similar products and trial and error by the manufacturer [2].

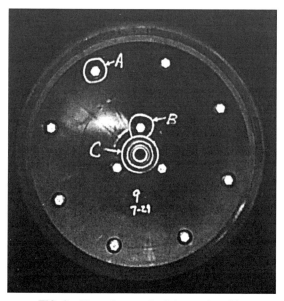

FIG. 2—*View of gage wheel from nylon side.*

Cracking occurred during assembly in three locations. (1) Along the outer bolt circle (marked A in Fig. 2), cracking of the nylon bolt seating during bolt installation sometimes occurs when an initial torque level of 200 in.-lb is used. (2) Along the inner bolt circle (marked B in Fig. 2), cracking occurs in the area where the bolt is seated. (3) Finally, cracking has occurred in the nylon around the bearing enclosure (that part of the nylon that touches the end of the bearing), distinguished by a loud, audible pop. The bearing retainer seat is marked C in Fig. 2.

In cases where audible pops were heard during the bolt installation and no visible cracks were found, it was assumed that the nylon had cracked internally. Both visibly cracked wheels and wheels assembled with audible pops were rejected at assembly. The rejection rate was considerably higher than the manufacturer could tolerate.

The manufacturer initially questioned the quality of the nylon or its processing. In addition, the manufacturer thought that the assembly loads might be too high for the nylon. Another concern was the possibility of large scatter in the nut torque, the torque required to drive the locknut onto a bolt. Unfortunately, actual material property data, stress analysis, and fatigue analysis for the wheel were not available. At this point, the authors were consulted. Drawings and a small number of assembled wheels were provided. Both cracked and un-cracked wheels were provided.

Initial Investigation

The authors' initial investigation focused on material quality and strength of the nylon wheel half, the assembly process, and the stresses and deflections in the nylon. No stress analysis was conducted on the steel wheel half because the manufacturer had not experienced any problems therein. Both experimental testing and analysis of the nylon wheel half, shown in Fig. 3, were performed in the initial investigation.

Experimental Testing

The authors' initial investigation included five major components: (1) sectioning of several wheels that had been rejected after assembly, (2) the determination of failure loads for the nylon bearing housing, (3) the determination of ultimate strength of the as-molded nylon, (4) the statistical determination of nut torque, and (5) torque relaxation in an assembled wheel.

The sectioning of several rejected wheels revealed no evidence of inferior molding. The visible cracks in bolt seats appeared to be the result of overtorques. A torque wrench was instrumented with strain gages and used to measure the torque required to untighten the previously assembled wheels. The bolts at the crack locations were obviously tightened beyond the 200 in.-lb level. Although a torque-limiting device was installed on the torque gun used in the assembly process, several of the bolts at cracked holes registered torque levels over 300 in.-lb during untightening. Therefore, the torque guns used in assembly were identified for future investigation.

The failure loads for the nylon bearing seat were determined by applying a compressive load to the bearing large enough to punch the bearing through the nylon seat. Five nylon wheel halves, rejected because of the popping sounds during assembly and without visible damage, were tested. The average load to punch out the bearing was 1800 lb. The smallest load was 900 lb and the largest load was 2600 lb. In addition, three nylon wheel halves, previously unassembled, were tested. The load to punch out the bearing was 3100 lb with no appreciably scatter. Therefore, the interference fit used in the assembly was identified for future investigation.

FIG. 3—*Nylon wheel half: (mating side to steel wheel).*

Eight tensile specimens were cut from the ribs of a single nylon wheel. These specimens were approximately 3 in. long with a width of either 0.3 or 0.6 in. and a thickness of 0.1 in. The specimens were tested in a screw-driven universal tensile testing machine. Failure loads were recorded and the ultimate tensile strength of each specimens was calculated by dividing the failure load by the cross-sectional area of the specimen. The average ultimate tensile strength was 9300 psi (64 123 kPa) and the standard deviation was 900 psi (6205 kPa). A lowerbound value of tensile strength was determined by subtracting three standard deviations from the mean. Thus, a lowerbound tensile strength of 6700 psi was established. This tensile strength is typical for nylon [3]. Although the tensile tests were nonstandard tests, they were deemed adequate for the current problem. No future investigation of strength or fracture toughness was considered necessary.

The locknut torque of both cadmium and zinc nut and bolt combinations was experimentally determined. To measure the locknut torque, a bolt head was gripped in a vise and a nut was threaded onto the bolt using an instrumented torque wrench. The highest torque measured during the first 1/4 to 3/8-in. thread engagement was recorded. Approximately 40 each of cadmium and zinc nuts were tested in this manner.

For the cadmium nuts, the average locknut torque was 22.1 in.-lb with a standard deviation of 2.65 in.-lb. To ensure that 99.1% of all the nuts in the population (not just the test set) were accounted for in later analysis, an upperbound nut torque was calculated by adding three standard deviations to the mean value. Thus, the upperbound locknut torque for the cadmium nuts was 30 in.-lb.

For the zinc nuts, the average locknut torque was 21.2 in.-lb and the standard deviation was 3.75 in.-lb. The upperbound locknut torque was determined in the same fashion for the zinc as was previously discussed for the cadmium nuts. The upperbound locknut torque for the zinc nuts was 31.5 in.-lb. Clearly, there is little difference between the locknut torque of cadmium and zinc nut and bolt combinations.

Using an instrumented torque wrench, the investigators hand-assembled eight wheels using 200 in.-lb of torque. The bolt-head torque was measured at times varying from 30 min to 24 h after assembly. The rate of torque relaxation for the first four hours is very high. After 16 h, the torque has essentially reached a steady state, average value between 140 and 145 in.-lb as seen in Fig. 4. Essentially, the torque relaxed to 65% of its original value in 24 h.

Torque values after 24 h were measured for a second set of hand-assembled wheels. The mean torque value was 159 in.-lb with a standard deviation of 6.76 in.-lb. A convenient lowerbound torque value encompassing 99.1% of the data can be calculated by subtracting three standard deviations from the mean value. Thus, a lowerbound torque of 139 in.-lb was determined.

During the assembly process, the authors did not experience any nylon cracking in the bolt seat areas (regions A and B in Fig. 2). However, audible pops were noted and assumed to be indicative of possible cracking in the nylon bearing seat.

Initial Analysis

In the current design, the nylon and steel wheel halves form a clamped retainer for the wheel bearing (Fig. 5). The current design uses in interference fit to firmly hold the wheel bearing between the two wheel halves. The interference was assumed to be the difference between the bearing length, L_b, and the cavity between a wheel assembled without a bearing, L_c. The amount of interference was determined by assembling a wheel without a bearing and measuring the gap between the nylon bearing seat and the steel bearing seat. A typical value for interference was 0.040 in.

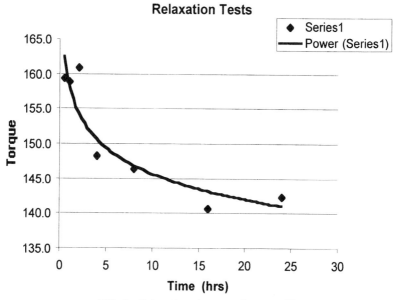

FIG. 4—*Relaxation of torque after assembly.*

Nylon

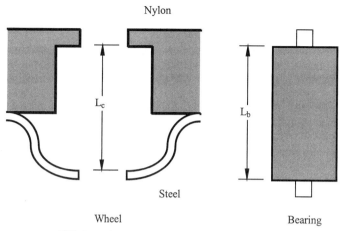

Wheel   Bearing

FIG. 5—*Schematic of bearing seats and bearings.*

The interference fit creates an axial compressive force on the bearing. This compressive force causes deflection and stress in the steel and nylon. Assuming that the bearing has a high axial stiffness, the interference fit is accommodated by deflection in both the steel and nylon. The sum of the deflections in the steel and nylon equals the amount of interference. Further, the force generated from interference during assembly is felt equally by both the steel and nylon.

A simple mechanics of materials model was solved for the force on the bearing for an interference of 0.040 in. A 10-deg sector of the steel wheel was treated as a straight, cantilevered beam. The force P applied to the steel is $1/36$ of the total interference force F_{ax}. The deflection in the steel, v_{steel}, is given by

$$v_{steel} = \frac{PL^3}{3EI}$$

where $E = 30 \times 10^6$ psi, $L = 0.82$ in., and $I = 1.322 \times 10^{-5}$ in.[4]. Substituting these values yields $v_{steel} = 1.287 \times 10^{-5} F_{ax}$. The deflection on a 10-deg section of the nylon was determined using the same beam formula with $E = 0.3 \times 10^6$ psi, $L = 0.1115$ in., and $I = 4.685 \times 10^{-5}$ in.[4]. The resulting expression for the deflection in the nylon is $v_{nylon} = 9.132 \times 10^{-7} F_{ax}$. The total deflection due to the bearing interference, v_{total}, is

$$v_{total} = v_{steel} + v_{nylon} = 0.040 \text{ in.}$$

The previously calculated deflections in terms of F_{ax} were substituted into the above equation and solved to obtain $F_{ax} = 2900$ lb. This value of F_{ax} is very close to the experimentally determined failure load of 3100 lb in the nylon bearing seat. Thus, the typical wheel would be assumed to be near the point of failure at assembly. Certainly, a redesign of the interference fit is required.

The Redesign

Load Analysis

A load analysis of the planter wheel is performed to determine the maximum loads that are to be carried through the bolted connections and the expected loads at the bearing. In finding the maximum loading on the bolted connections, several worst-case assumptions are made. For example, the entire load carried by the wheel is assumed to be transmitted from the steel to the nylon through the bolted connections. Also, while two gage wheels are mounted on each planter row unit, one wheel is assumed to carry the planter row unit weight of 400 lb. A free-body diagram of the planter wheel with applied loads is shown in Fig. 6.

In Fig. 6, there are three basic forces assumed to be applied to the wheel in a worst-case loading situation. First, a normal force (equal to the load weight) causes both radial and shear loads due to an offset angle θ, (approximately 12 deg) of the gage wheels. This load is applied at the extreme edge of the wheel. Second, a transverse force applied at the edge of the wheel causes radial and shear loads. This load would result if the planter units were in the ground while making a turn with the planter, or with operation on a slope. This force is assumed to equal a factor, β, of the load on the planter. Finally, a force on the leading edge of the wheel, at approximately $(1-\cos(\alpha))$ of the height of the wheel, results in radial loads on the wheel.

As a final note on the loading, all three loads also cause bending moments at the bearing about the y and z axes (x axis aligned with the bearing axis).

Translating these applied loads to point O, the point where the center of the bearing is supported by the nylon retaining seat (see Fig. 6) gives the following loads

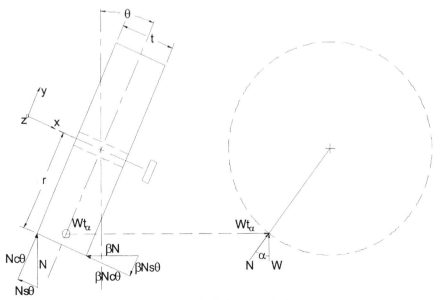

FIG. 6—*Free-body diagram of wheel.*

$$M_y = -W \tan(\alpha)\, t/2$$

$$M_z = -[W \cos(\theta) - \beta\, W \sin(\theta))t + (W \sin(\theta) + \beta\, W \cos(\theta))r]$$

$$F_x = -(W \sin(\theta) + \beta\, W \cos(\theta))$$

$$F_y = (W \cos(\theta) - \beta\, W \sin(\theta))$$

$$F_z = W \tan(\alpha)$$

Given the loading conditions

$$W = 400 \text{ lb}, \theta = 12 \text{ deg}, \alpha = 30 \text{ deg},$$

$$\beta = 0.5, r = 7.5 \text{ in.}, t = 4 \text{ in.}$$

$$M_y = -461 \text{ in.-lb}$$

$$M_z = -3490 \text{ in.-lb}$$

$$F_x = -279 \text{ lb}$$

$$F_y = 350 \text{ lb}$$

$$F_z = 231 \text{ lb}$$

From these loads in the x-y-z coordinate system, the radial and axial loads on the wheel and bearing are

$$F_r = (F_y^2 + F_z^2)^{1/2} = 419 \text{ lb}$$

$$F_{ax} = F_x = -279 \text{ lb}$$

Now, examine the loading at the outer bolt ring. Assume first that the 8 bolts in the outer ring carry all radial, axial, and bending loads. The bending loads, M_y and M_z can be replaced with a couple acting at the outer bolt ring (Fig. 7). In this case, only M_z will be considering since it is the dominant bending.

$$C = C' = 0.5(M_z/r_b) = 323 \text{ lb}$$

Consider the load carried from ground to steel to nylon to the bearing. The bolts transmit the load from the nylon to steel. Consider the eight bolts in the outer ring to carry the entire load. Then, the direct axial load on a bolt is the sum of the total axial load, divided by eight, plus the bending couple, C, which is carried by two bolts

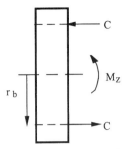

FIG. 7—*Schematic of force-couple to resist bending about z-axis.*

$$F_a = F_{ax}/8 + C/2 = 196 \text{ lb}$$

Assume the shear load is carried by friction between the nylon and steel. This requires a normal force in the bolt

$$F_n = (F_r/\mu)/8 = 210 \text{ lb}$$

Thus, the maximum tension in the bolts required to carry both axial and shear loads is

$$F_b = F_a + F_n = 406 \text{ lb}$$

The desired preload specified in the bolted connection is set equal to the maximum load to be carried by the bolted connection. This insures a constant state of compressive stress in the nylon member and will limit the fatigue loading.

The desired interference (difference in bearing and space length) for the wheel assembly is calculated based on the desired compressive axial holding load to be maintained on the bearing. The maximum axial load carried by the wheel bearing is equivalent to the maximum axial load on the wheel calculated above

$$F_{ax} = 279 \text{ lb}$$

A 558-lb compressive force on the bearing would exceed the maximum expected axial load by a factor of two. This will guarantee the bearing to remain under compression during operation with a factor of safety of two.

The interference required for this axial load

$$y_{inter} = (1.291 + 0.09131) \times 10^{-5} F_{ax} = 1.382 \times 10^{-5} (558) = 0.0075 \text{ in.}$$

In summary, an interference of approximately 0.007 to 0.008 in. will keep the bearing in compression during use.

Torque Analysis

The required torque for the bolts is calculated based on the desired bolt preload. The required torque has three components: torque needed for preload, torque needed to overcome locknut friction, and torque needed to overcome bolt-head friction (since input torque is applied to the bolt head).

The torque required to achieve preload, T_p, is linearly related to the bolt preload F_b

$$T_p = K d F_b$$

where K is the torque coefficient and d is the diameter of the bolt. The torque coefficient is a function of the bolt condition, i.e., whether the bolt is nonplated, zinc- or cadmium-plated, or lubricated. Shigley [4] gives a typical value of $K = 0.16$ for cadmium-plated bolts. For the preload of 406 lb and bolt diameter of 0.375 in., the preload torque is calculated to be 24.4 in.-lb.

The torque required to overcome locknut friction, T_{in}, was statistically determined as the upperbound torque (mean torque plus three standard deviations) measured in the initial testing program

$$T_{in} = T_{mean} + 3\ T_{sd}$$

For the cadmium nuts, T_{mean} was 22.1 in.-lb and T_{sd} was 2.65 in.-lb. The resulting T_{in} was calculated to be 30.1 in.-lb.

The torque required to overcome bolt-head friction, T_{bh}, is linearly related to the preload

$$T_{bh} = \mu\ \frac{d + d_0}{4}\ F_b$$

where μ is the coefficient of friction, d is the bolt diameter, and d_0 is the bolt head diameter. In this analysis, μ is assumed to be 0.25. For $d_0 = 0.5$ in., T_{bh} was calculated to be 22.4 in.-lb.

The required torque, as measured from the bolt-head is the sum of the three component torques (preload, locknut friction, and bolt-head friction)

$$T_{reqd} = T_p + T_{in} + T_{bh}$$

Using the previously calculated and measured torques yielded $T_{reqd} = 77$ in.-lb. However, since the torque relaxes significantly after assembly, the initial assembly torque, T_i, must be substantially larger than the torque required for service. Assuming that the torque will relax to 65% of its initial value, as observed in the initial testing, the initial assembly torque must be 100/65 of the required torque or $T_i = 120$ in.-lb.

Static Factor of Safety for Nylon Bolt Seat

The minimum cross section of nylon in the bolt seat is cylindrical with inside diameter $d_i = 0.338$ in. and outside diameter $d_0 = 0.725$ in. The maximum stress in the nylon bolt seat is the axial stress given by

$$\sigma_{max} = \frac{F_b}{\frac{\pi}{4}\ (d_0^2 - d_i^2)}$$

Using $F_b = -406$ lb leads to $\sigma_{max} = -890$ psi. The static factor of safety, N_s, is given by

$$N_s = \frac{S_{ut}}{|\sigma_{max}|}$$

Using $S_{ut} = 6700$ psi leads to $N_s = 7.5$.

Fatigue Factor of Safety for Nylon Bolt Seat

The fatigue life is highly dependent on the alternating and mean values of applied stress or equivalently, applied force. The resultant force in the nylon member, F_m, is

$$F_m = \frac{k_m}{k_b + k_m} P - F_b$$

where k_m is the stiffness of the nylon member, k_b is the stiffness of the bolt, P is the total applied external force on the bolted joint, and F_b is the force on the bolt. The stiffnesses were determined by mechanics of materials methods outlined by Shigley [4] to be $k_m = 0.089 \times 10^6$ lb/in. and $k_b = 1.7 \times 10^6$ lb/in. Using $P = 196$ lb and $F_b = 279$ lb results in a member force $F_m = -270$ lb. Assuming that P varies from zero to its maximum value of 196 lb, the minimum value of F_m is -279 lb and the maximum value is -270 lb. Thus, the nylon member is always loaded in compression. The alternating stress, F_{lat}, is 5 lb and the mean stress, M_{ean}, is -275 lb. Since the mean stress is compressive, the factor of safety against fatigue failure, N_f, is given by

$$N_f = \frac{S_e}{\sigma_a}$$

where S_e is the fatigue endurance limit and σ_a is the alternating stress (calculated in the same manner as σ_{max} in the static analysis). The endurance limit for nylon was estimated to be 20% of the ultimate strength: $S_e = 1300$ psi. The alternating stress, σ_a, was calculated to be 10 psi. The fatigue factor of safety, N_f, was calculated to be 130.

Assembly Tests

The investigators went to the manufacturing facility to conduct a series of assembly tests to validate the redesign. The torque gun used in the assembly was calibrated using a calibration system typical of the aerospace industry. The assembly torque was reduced from 200 to 120 in.-lb and approximately fifty wheels were assembled. No cracking of bolt seats occurred using the redesign torque level. However, several wheels were assembled with audible pops. Even with the reduced torque levels, several nylon hubs were damaged. The failures were not unexpected because the assembled wheels had not been reworked to reduce the interference fit.

Bearing Interference Analysis

Even with the reduced torque levels designated in the redesign, nylon cracking occurred in the bearing seat, as evidenced by audible pops in approximately 5% of the wheels. Physical investigation led to a bearing interference analysis. First, the bearing interference on the current wheel design was measured at the torque levels specified in the new design.

This interference was found to create an axial force of 2900 lb on the bearing, far greater than that needed. Referring to the original load analysis, an axial bearing force of 560 lb was prescribed (this was twice the maximum expected axial load). This required an interference of 0.007 to 0.008 in. in the axial bearing direction and would greatly lower stresses in the nylon wheel around the rim of the bearing retainer. Corrections for this newly specified bearing interference could be made either in the steel side stamping or the nylon side molding.

Discussion

Based on the redesign, the following recommendations were made to the manufacturer. The first recommendation was to reduce the assembly torque from 200 to 120 in.-lb. The second recommendation was to reduce the bearing interference from 0.035 to 0.040 in. to approximately 0.007 to 0.008 in. Based on the capabilities of their suppliers, the manufacturer chose to modify the steel stamping. To expend the current stock of steel wheel halves, it was concluded that the inner bolt circle could be assembled with an initial torque of 60 in.-lb, relaxing to 41 in.-lb. The outer bolt circle would be assembled with an initial torque of 120 in.-lb, relaxing to 77 in.-lb. Even with this reduced torque level on the inner bolts, a statistical sampling of the wheels was recommended. A final recommendation was to begin a regular maintenance schedule for torque guns used in the assembly process.

Conclusions

As a final note for this work, it is important to reflect on the role of this paper in a symposium on the structural integrity of fasteners. The authors make two important observations. The first observation is that a bolted joint consists of two major components: the fastener system and the material being joined. The title of this symposium rightly suggests that the structural integrity of the fastener is of paramount importance. However, this paper is a reminder that the joined material can also fail. As in most cases, it is the weakest link that must be critically examined. The second observation is that cracking problems are not always solved by the application of fracture mechanics. Many cracking problems actually have their underlying cause with inadequate machine design. These two observations should serve as useful reminders to both designers and analysts.

References

[1] Ace Products, Inc., "Ace Introduces Four New Wheel Assemblies to Increase Crop Yields," Ace Products, Inc., Newport, TN, 1997.
[2] Telephone conversation with Gary Moore of Ace Products, Inc., Newport, TN, April 17, 1998.
[3] Callister, W. D., Jr., *Materials Science and Engineering: An Introduction,* 4th ed., Wiley, New York, 1997.
[4] Shigley, J. E. and Mischke, C. R., *Mechanical Engineering Design,* 5th ed., McGraw Hill, New York, 1989.

Victor K. Champagne[1]

Failure Analysis of High Strength Steel Army Tank Recoil Mechanism Bolts

REFERENCE: Champagne, V. K., "**Failure Analysis of High Strength Steel Army Tank Recoil Mechanism Bolts**," *Structural Integrity of Fasteners: Second Volume, ASTM STP 1391,* P. M. Toor, Ed., American Society for Testing and Materials, West Conshohocken, PA, 2000, pp. 48–62.

ABSTRACT: A failure analysis of two broken bolts from an Army tank recoil mechanism was performed by the U.S. Army Research Laboratory (ARL), Weapons and Materials Research Directorate. The bolts failed at the head-to-shank radius during installation at Aniston Army Depot. A total of 69 additional bolts from inventory and the field were also characterized and tested for comparison. Optical and electron microscopy of the broken bolts showed topographies and black oxide on the fracture surface consistent with the characteristics of quench cracks. The crack origin was located within a region covered with a heavy black oxide where fracture occurred as a result of intergranular decohesion. Heat treatment tests of the material confirmed the black oxide to be formed during the tempering operation of ~677°C. The remaining fracture surface failed in a ductile fashion. The cause of failure was attributed to preexisting quench cracks which should have been detected by the required 100% magnetic particle inspection conducted during manufacturing. These cracks propagated during installation, causing the bolt heads to sever. Recommendations were provided to screen inventory to prevent cracked parts from entering fielded tanks and improve control of manufacturing and inspection procedures. These included machining the head-to-shank radii to specification, tighter control of the magnetic particle inspection process, the use of dull cadmium plate to mitigate the potential for delayed failures due to hydrogen embrittlement or stress corrosion cracking, an alternative to electrolytic cadmium plate (such as vacuum deposition), and finally, replacing all existing bolts in the field and in inventory with new or reinspected bolts.

KEYWORDS: failure analysis, fractography, quench cracks, AISI 8740 steel

A comprehensive metallurgical analysis was performed by the U.S. Army Research Laboratory (ARL) to determine the cause of failure of two cadmium plated steel bolts (P/N 9328588–4, -3) from an M60 tank A3 recoil mechanism. These two bolts (Bolt #1 and Bolt #2) fractured at the head-to-shank radius during installation. Preexisting quench cracks appear to have been the major contributor to the failure.

Approach

A total of 69 additional bolts were received from the Aniston Army Depot and the Army Material Command for comparative examination and testing by ARL. These bolts were obtained from both the inventory and the field.

The 71 bolts were magnetic particle inspected for cracks. Subsequently, the following analyses and tests were performed: chemical composition of the alloy; measurement of the

---

[1] Chief, Materials Analysis Group, U.S. Army Research Laboratory, Weapons and Materials Research Directorate, Aberdeen Proving Ground, MD 21005-5069.

radius at the bolt shoulder/shank interface for indication of excessive stress concentration; mechanical properties and hardness measurements; metallographic examination for microstructural characterization of the alloy and cadmium plating thickness and uniformity; torque testing for maximum torque-to-failure; stress durability testing for externally threaded fasteners which may be subject to any type of embrittlement (such as hydrogen embrittlement induced by cadmium electroplating); scanning electron microscopic examination of fracture surfaces; and elevated temperature exposure tests at the tempering temperature (~650°F [343°C]) and stress relief temperature (~190°F [88°C]) to determine when, during the processing cycle, the observed high temperature oxide occurred.

Table 1 contains a listing of the bolts received, the size, the radius at the shoulder and shank interface and the type of test or examination performed. Abbreviations have been used for the following terms: metallography—MET, hardness—HARD, fractography—FRAC, energy dispersive spectroscopy—EDS, and magnetic particle inspection—MPI.

Results

Chemical Analysis

The drawing and specifications for the bolt allow the fastener to be fabricated from any of the following steels: AISI 4140, 4340, 6150, or 8740. Atomic absorption and inductively coupled argon plasma emission spectroscopy were used to determine the chemical composition of two of the bolts. Carbon and sulfur contents were determined by the LECO combustion method. The chemical analysis revealed that the bolts were fabricated from AISI 8740 steel (Table 2). This low alloy steel is quite similar in properties to AISI 4130. In the quenched and tempered conditions, the alloy exhibits a good combination of strength, toughness, and fatigue resistance.

Radius Measurement at Head-to-Shank Radius

The radius requirement at the head-to-shank interface was 0.145, +0.0000, −0.0025 cm. Approximately 50% of the 59 bolts measured did not meet the specification requirement. The radii of these bolts were sharper than specified and have the potential to provide sites for crack initiation due to higher stress concentrations.

Microstructural Analysis

Longitudinal and transverse metallographic samples taken from the failed bolts were prepared. In the as-polished condition, the material appeared clean, with no evidence of major inclusions or inherent internal defects. Once etched with a 2% nital solution, the microstructure of both bolts consisted of a tempered martensitic structure, typical of a quenched and tempered low alloy steel.

Cadmium Plating Thickness and Uniformity

The bolts were required to be cadmium electroplated in accordance with Federal Specification QQ-P-416, Type II, Class 2 (0.0076-mm thickness minimum). The average thickness of the cadmium plating of Bolt #1 was 0.012 mm, and the plate was quite uniform. The plating on Bolt #2 was much thinner (0.004 mm), and less uniform. In addition, a bolt received from inventory was chosen at random and metallographically prepared. Table 3 shows that this bolt exhibited a uniform cadmium plating thickness of 0.010 mm. According

TABLE 1—*Bolt tabulation data.*

Bolt No.	Size	Radius, cm	Work Performed
1	4	Broken	MET, HARD, FRAC, EDS, MPI
2	4	Broken	MET, HARD, FRAC, EDS, MPI
3	4	0.122	Chem. Anal.
4	4	0.122	Chem. Anal.
5	4	0.122	MET, HARD, FRAC, MPI
6	4	0.122	MET, HARD, FRAC, MPI
7	4	0.122	MET, HARD, FRAC, MPI
8	4	0.122	Stress Durability, MPI
9	4	0.122	Stress Durability, MPI
10	4	0.127	Stress Durability, MPI
11	4	0.127	Stress Durability, MPI
12	4	0.132	Stress Durability, MPI
13	4	0.132	Tensile Test (Failed)
14	4	0.142	MPI
15	4	0.119	MPI
16	4	0.122	MPI
17	4	0.122	MPI
18	4	0.122	MPI
19	4	0.122	MPI
20	4	0.122	MPI
21	4	0.124	MPI
22	4	0.127	MPI
23	4	0.127	MPI
24	4	0.127	MPI
25	4	0.130	MPI
26	4	0.130	MPI
27	4	0.132	MPI
28	4	0.132	MPI
29	4	0.132	MPI
30	4	0.140	MPI
31	4	0.142	Stress Durability, MPI
32	4	0.152	MET, HARD, FRAC, EDS, MPI
33	4	0.142	Stress durability, MPI
34	4	0.145	MPI
35	4	0.122	MPI
36	4	0.135	MET, HARD, FRAC, EDS, MPI
37	4	0.145	MET, HARD, FRAC, EDS, MPI
38	4	0.147	MET, HARD, FRAC, EDS, MPI
39	4	0.147	Stress Durability, MPI
40	4	0.147	Stress Durability, MPI
41	4	0.147	Stress Durability, MPI
42	4	0.147	Stress Durability, MPI
43	4	0.147	Stress Durability, MPI
44	4	0.147	Stress Durability, MPI
45	4	0.147	MPI
46	4	0.147	MPI
47	4	0.147	MPI
48	4	0.147	MPI
49	3	0.150	MPI
50	3	0.142	MPI
51	3	0.142	MPI
52	3	0.142	MPI
53	3	0.147	MPI
54	3	0.147	MPI
55	3	0.147	MPI
56	3	0.147	MPI
57	3	0.147	MPI
58	3	0.147	MPI
59	3	0.150	MPI

TABLE 1—*Continued.*

From Ft. Knox			
1H	...	...	Tensile Test
1T	...	...	Tensile Test
2H	...	...	Tensile Test
2T	...	...	Tensile Test
3H	...	...	Tensile Test
3T	...	...	Tensile Test
4H	...	...	Tensile Test
4T	...	...	Tensile Test
5H	...	...	Tensile Test
5T	...	...	Tensile Test
6H	...	...	Tensile Test
6T	...	...	Tensile Test

NOTE—MET = Metallography.
FRAC = Fractography.
HARD = Hardness.
EDS = Energy dispersive spectroscopy.
MPI = Magnetic particle inspection.
H = Specimen machined from head portion of bolt.
T = Specimen machined from thread portion of bolt.

to these measurements, it was determined that Bolt #1 conformed to a Class 2 coating, while Bolt #2 compared more favorably in thickness to a Class 3 coating.

Magnetic Particle Inspection

Each bolt that was received from inventory and service was magnetic particle inspected for cracks and discontinuities in accordance with MIL-I-6868. One of the bolts exhibited the presence of a crack. Figure 1 shows the crack revealed by using black light photography. This bolt contained a transverse crack near the head-to-shank radius that extended over two-thirds of the circumference of the bolt shank. It was assumed that bolts had been previously 100% inspected in accordance with MIL-I-8831. Since this bolt failed the inspection, it

TABLE 2—*Chemical analysis (weight percent).*

Element	AISI 4140	AISI 4340	AISI 8740	AISI 6150	Bolt A	Bolt B
Aluminum	...	...	...	...	0.023	0.021
Chromium	0.8–1.1	0.7–0.9	0.4–0.6	0.95	0.52	0.53
Copper	0.35	0.35	0.35	...	0.12	0.03
Manganese	0.75–1.0*	0.6–0.9*	0.75–1.0*	0.75	0.82	0.88
Molybdenum	0.15–0.25*	0.2–0.3	0.15–0.25*	0.05	0.19	0.19
Nickel	0.25	1.65–2.0	0.4–0.7*	0.15	0.59	0.68
Phosphorus	0.025	0.015	0.025	0.025	0.015	0.007
Silicon	0.2–0.35*	0.15–0.25*	0.2–0.35*	0.25	0.21	0.27
Sulfur	0.02	0.015	0.04	0.025	0.01	0.006
Carbon	0.4*	0.38–0.43*	0.38–0.43*	0.5	0.36	0.43

*Values correspond with specimen values.
NOTE—Bolts correspond with specifications for AISI 8740 steel.

TABLE 3—*Cadmium plating thickness measurements.*

Bolt No.	Area # 1, mm	Area #2, mm	Area #3, mm	Area #4, mm	Average Thickness, mm
Failed No. 1	0.004	0.004	0.003	0.0045	0.00388
Failed No. 2	0.012	0.013	0.011	0.011	0.01175
Inventory	0.01	0.01	0.011	0.0095	0.01013

appears that either a 100% inspection was not carried out, or if it was, the defect was not detected by the inspector.

Mechanical Properties and Hardness Measurements

A standard ASTM tensile test was performed on six bolts which had experienced extensive field use at Ft. Knox (see Table 4). The specimens (0.287-cm in diameter) were fabricated from the bolt head and thread area. Also, a specimen was fabricated from the bolt from inventory that had exhibited the cracking noted in the "Magnetic Particle Inspection" section. Table 5 contains the mechanical property data obtained. In addition, tensile tests were performed on the new inventory bolts listed in Table 6. The actual bolts were tested in tension, and all failed in the threads with the exception of one which failed at the bolt head. The ultimate tensile strength (UTS) of these inventory bolts ranged from 1371 to 1467 MPa. It

FIG. 1—*Black light photograph of bolt which exhibited a quench crack after MPI. Magnification: approximately ×1.*

TABLE 4—*Bolt history.*

Bolt No.	Number of Rounds Fired
1	508.5
2	508.5
3	125.0
4	125.0
5	864.0
6	864.0
7	837.0
8	837.0
9	942.5
10	942.5
11	647.0
12	647.0

NOTE—Used bolts from Ft. Knox.

should be noted that only 1 out of the 11 bolts had a UTS below 1441 MPa. The maximum load-to-failure was between 19 391 and 20 752 kg, exceeding the minimum specified ultimate tensile load of 17 736 kg.

A Knoop hardness survey was conducted on six fielded bolts from Ft. Knox. The results and equivalent HRC values are listed in Table 7. Hardness measurements taken on two new bolts from inventory ranged from equivalent HRC values of 40 to 42 (Table 8). There was no evidence of a hardness gradient from surface to core on any of the bolts tested. The bolts were required to have a hardness of 39 to 43 HRC in accordance with MIL-B-8831. The new bolts met this requirement; however, bolts with a firing history taken from the field exhibited higher hardness values (44 to 45 HRC).

TABLE 5—*Tensile test results (specimens machined from bolts).*

Spec. ID	Area, cm²	0.1% Y.S., MPa	0.2% Y. S., MPa	UTS, MPa	% RA	% El
FS	0.0671	1322	1326	1409	53.9	12.2
1H	0.0639	1316	1330	1417	55.2	10.6
1T	0.0645	1379	1396	1503	52.0	12.0
2H	0.0645	1379	1396	1503	52.0	12.0
2T	0.0658	1311	1331	1453	52.0	11.0
3H	0.0645	1324	1338	1420	56.0	11.9
3T	0.0645	1311	1345	1420	53.0	12.1
4H	0.0658	1298	1305	1386	53.9	11.0
4T	0.0645	1258	1310	1413	56.0	12.7
5H	0.0658	1335	1352	1420	55.9	12.6
5T	0.0658	1318	1335	1420	55.9	12.6
6H	0.0671	1325	1359	1432	54.8	13.2
6T*	0.0658	...	...	...	...	

NOTES—Used bolts from Ft. Knox were tested.
FS = Specimen machined from quench cracked bolt.
H = Head portion of bolt.
T = Thread portion of bolt.
*Specimen was not tested due to improper machining.

TABLE 6—*Tensile test results (actual bolts).*

Bolt No. (see Table 1)	Max. Load, kg	UTS, MPa
9	20 457	1447
44	20 412	1443
43	20 752	1467
40	20 457	1447
32	20 639	1459
41	20 639	1459
10	20 639	1459
42	20 593	1456
11	20 412	1443
12	20 639	1459
13	19 391	1371

Torque Testing

Torque testing was performed on eight bolts from the field and from inventory employing a calibrated torque wrench equipped with a heavy-duty socket. The torque-to-failure was within 374 to 612 N-m as shown in Table 9. These values exceeded the minimum torque requirement of 163 to 190 N-m. Seven of the bolts failed at the beginning of the threaded section while the remaining bolt failed in the center of the threaded region. Since none of the bolts failed at the head-to-shank radius, the measured radius at this interface did not contribute to the failure. Generally, the bolts from inventory exhibited higher torque failure values when compared to the bolts from the field.

Stress Durability Test

To investigate the possibility that hydrogen may have been introduced into the bolt during the cadmium plating operation and may not have been adequately removed and/or dispersed by the low temperature embrittlement relief treatment (causing hydrogen embrittlement), a stress durability test was performed in accordance with MIL-STD-1312-5A, Test 5. A plate fixture was fabricated from AISI 4140 steel and heat treated to 45 HRC. This plate was

TABLE 7—*Knoop microhardness survey (500 g major load, diamond penetrator).*

Spec. ID	Knoop	Eq. HRC
1H	472	45.5
1T	459	44.5
2H	457	44.0
2T	447	43.0
3H	458	44.5
3T	454	44.0
4H	461	44.5
4T	467	45.1
5H	466	45.0
5T	467	45.2
6H	466	45.0
6T*	461	44.5

TABLE 8—*Rockwell "C" scale profile (HRC) on two new bolts.*

	Transverse	Longitudinal
Bolt With Quench Crack	42.8 HRC	41.3 HRC
	41.3	43.1
	43.0	43.3
	42.2	42.9
	41.6	41.9
Average HRC	42.2	42.5
Total Average	42.3	
Bolt No. 5 (see Table 1)	38.5	46.3
	39.8	40.7
	42.3	43.4
	40.6	36.0
	40.6	36.6
Average	40.4	40.6
Total average	40.5	

drilled and tapped to accommodate 16 bolts. Sixteen load cells were also fabricated from AISI 4140 steel, 45 HRC. The load cells were strain gaged and calibrated for load versus strain. Eleven new bolts and six used and fielded bolts were preloaded to 80% of the UTS and subjected to a 200-h test. None of the bolts fractured as a result of this testing. Optical examination of the bolts subsequent to testing did not reveal the presence of any transverse cracking. Note that MIL-B-8831 specifies that the preloaded bolt shall be maintained at load for only 23 h without failure. In order to provide a better statistical sampling of the bolt inventory, 12 additional new bolts were tested, and the duration of the test was extended to 400 h. Again, there were no failures after stressing these bolts at 80% UTS for 400 h.

Fractographic Examination

The fracture surfaces of the two failed bolts were subsequently examined in the scanning electron microscope (SEM). Figure 2 contains a macrograph of the fracture surface of Bolt #1 obtained by light optical microscopy. There were four distinct fracture zones, which are displayed schematically. A black covered area at the periphery of the circular fracture surface was observed and labeled Zone 1. River markings indicative of crack growth were also observed. Adjacent to Zone 1 was a gray area, Zone 2, which also contained river markings.

TABLE 9—*Torque test results.*

Bolt No. (see Table 1)	Radius, cm	Prior History	Failure Torque, N-m
16	0.122	Inventory	374
17	0.122	Inventory	612
18	0.122	Inventory	605
27	0.132	Inventory	598
46	0.147	Field	408
47	0.147	Field	476
48	0.147	Field	428
50	0.150	Field	571

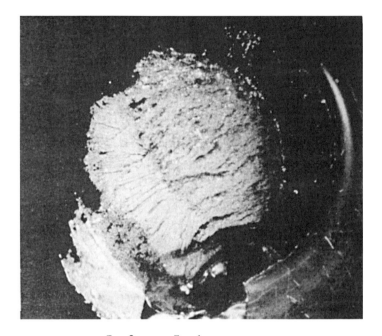

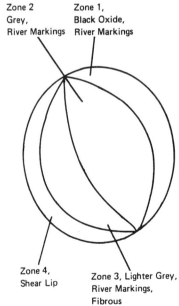

FIG. 2—*Fractured heat of Bolt #2 showing different zones of failure. Magnification: ×5.*

Another light gray area, Zone 3, was similar to the Zone 2 gray area and was observed adjacent to a small shear lip around the edge of the bolt (Zone 4). Metallographic examination had previously shown that the structure consisted of tempered martensite, and no internal defects were noted. Scanning electron microscopic examination within Zone 1 revealed an intergranular fracture surface below the black layer observed visually. Energy dispersive spectroscopy (EDS) of this surface showed those elements associated with the steel, as well as oxygen (Fig. 3). The black material was concluded to have been an oxide and not a contaminant. There was, however, one small inclusion found with high levels of silicon and cadmium. The cadmium apparently came from the plating. Generally, the material was inclusion free. Zone 2 was characterized by a mixed intergranular and ductile dimpled morphology, while Zone 3 contained this mixed type of fracture mode in a very fibrous manner. There appeared to be more doctile dimpling in Zone 3 than in Zone 2. The fracture surface of Bolt #2 exhibited the same features as those described above; Zone 1 was intergranular with a covering of black oxide; Zones 2 and 3 showed a mixed morphology of intergranular and ductile dimpling, but with more ductile fracture, and Zone 4 was the typical shear/fast fracture surface.

The fracture surface of the bolt which was found to contain a quench crack was also examined. This fracture was noticeably different than the other two failed bolts. The entire surface of the fracture was covered with the black oxide except for a narrow shear lip shown in Fig. 4. The same fibrous zones and features such as river patterns were apparent macroscopically, but not as markedly. The major differences were observed in the SEM. All the first three zones appeared to have the same intergranular fracture mode. There was no evidence of a ductile dimpled morphology as a result of tensile overload. The low tensile load of the bolt (approximately 454 kg) obtained during tensile testing, in conjunction with the

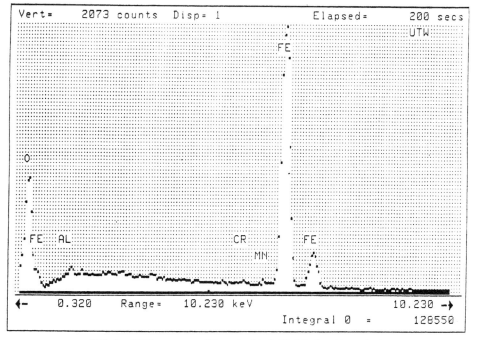

FIG. 3—*EDS spectrum of Zone 1 of bolt #2 (black oxide region).*

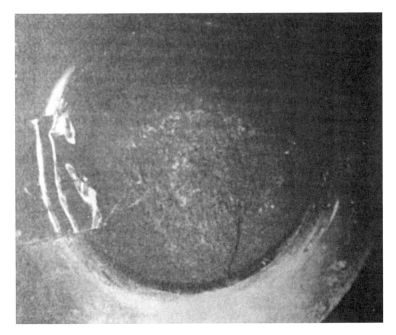

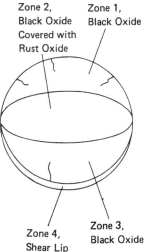

Zone 2,
Black Oxide
Covered with
Rust Oxide

Zone 1,
Black Oxide

Zone 4,
Shear Lip

Zone 3,
Black Oxide

FIG. 4—*Zones 1 and 3 were all considered Zone 1 since they were all intergranular fractures covered with a black oxide. Magnification: ×3.75.*

black oxide covering over 90% of the surface and the absence of ductile dimpled rupture, implies the crack area of the bolt encompassed the black oxide surface before tensile testing. This suggests that the cracks in both failed bolts which were also covered with black oxide, were pre-existing flaws and not due to the service environment. It appeared, therefore, that hydrogen embrittlement may be ruled out as a failure mode. In order to determine when the crack occurred during the bolt fabrication process, and when the black oxide film formed on the crack surface, elevated temperature tests were carried out. Figure 5 shows that there is substantial thickness of oxide on the fracture surface of the failed bolts. From Fig. 6, it appears that the oxide is beginning to diffuse into the base alloy. Disk specimens were sectioned from both the failed bolts, polished through 600 grit SiC paper, and cleaned. One specimen was placed into an oven preheated to 191°C, typical of a low temperature stress relief treatment, and another exposed to 649°C. The specimen heated to 191°C did not oxidize after exposure for 1 h. However, after exposure to 649°C for 1 h, the specimen was covered with black oxide. Indeed, the black oxide was present after only 5 min of exposure at this temperature. Considering that the bolts were heat treated at 871°C, quenched, tempered at 677°C, stress relieved at 191°C, cadmium plated and baked at 191°C (± 14°C) for 3 h to prevent hydrogen embrittlement, the cracks most likely occurred during quenching and the black oxide film formed during the tempering operation. Of the 12 bolts that were tensile tested as described earlier, the two which exhibited the lowest UTS were subjected to fractographic analysis. The fracture surfaces showed a mixed intergranular and dimpled rupture morphology and a fibrous texture similar to Zones 2 and 3 of the failed bolts. There was no evidence of a black oxide film. Of the eight inventory and field bolts that were torqued to failure (described earlier), representative bolts from inventory and the field were chosen for fractographic analysis. Shear dimpling was prevalent as expected in torque failures. The cup structure of shear dimpling was found throughout the fracture surfaces except for the 45° shear planes. All the bolts subjected to torque testing exceeded the minimum requirements for minimum torque-to-failure by 2 to 3 times.

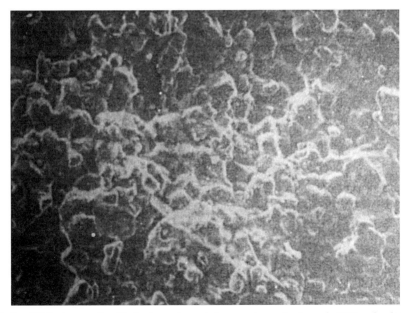

FIG. 5—*SEM micrograph of bolt shown to contain quench cracks through MPI and subsequently tensile tested. The surface exhibited a predominantly intergranular morphology covered with black oxide. Magnification: ×300.*

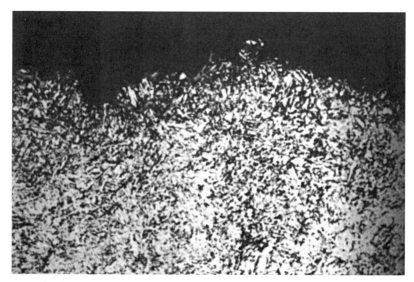

FIG. 6—*Microstructure and fracture surface of Bolt #2. Magnification: ×840.*

Discussion

Quench Cracks

Quench cracks in steels result from stresses produced during the austenite-to-martensite transformation, which is accompanied by an increase in volume. The observed cracks in the failed bolts meet the following characteristics of quench cracks [1]: the crack traverses from the surface toward the center of mass, grows and exhibits a shear lip at the outer surface; the crack does not exhibit any decarburization in a microscopic examination; when tempering after quenching, the fracture surface is blackened by oxidation. As previously described, a tempering scale (oxide) had formed in the quench cracks after tempering at higher temperatures under oxidizing conditions. Any condition that concentrates stresses that occur in quenching, promotes the formation of quench cracks. Distribution of mass and lack of uniform or concentric cooling of the part may promote cracking. In addition, selection of an unsuitable quenching medium may also be contributory. After quenching, the part should be tempered sufficiently to relieve the internal stresses formed in quenching (temper the part while it is still warm, i.e., 66 to 93°C as withdrawn from the quenching medium).

Further support of a quench crack fracture mechanism may be found in examination of both light and electron microscopic fractographs of typical quench cracks in an AISI 4340 steel [2]. These fractographs show the quench crack crescent where the crack is intergranular in nature. Comparable fractographs of the two failed bolts show the same features; the quench crack crescent designated Zone 1, and the intergranular fracture mode in Zone 1. Fractographic examination of the bolt found to contain cracking as a result of MPI also supports the contention that the cracks were preexisting quench cracks and not due to the service environment.

Stress Corrosion Cracking, Hydrogen Embrittlement

High hardness steels can fracture under very low static stresses if they are embrittled by hydrogen absorption or exposed to an environment capable of causing stress corrosion crack-

ing (SCC). Hydrogen embrittlement (HE) fractures frequently result from hydrogen perme-
ation into a metal during electroplating and can be difficult to extinguish from SCC fractures,
particularly when the environment is also a source of hydrogen. Both mechanisms usually
result in faceted, intergranular fracture origins in low alloy steels.

Hydrogen produced during cadmium plating can lead to catastrophic failure of a stressed
structural part. These plated parts must be baked to remove/disperse the hydrogen, lowering
the internal concentration and thus reduce possibility of failure. Conventional bright cadmium
deposited from cyanide baths is preferred due to its appearance and protective characteristics.
However, this plating is a barrier to hydrogen diffusion, and even prolonged baking may not
drive off all of the hydrogen. The degree of embrittlement becomes more severe with increasing steel strength. For example, AISI 4340 steel, 1793 to 1931 MPa UTS, with an acute
notch (K_t = 5.6) might be embrittled with less than 0.1 ppm mobile hydrogen. Therefore,
low embrittlement baths are used for high strength steels. These produce duller and more
porous plates which lose hydrogen more readily upon baking. However, the dull cadmium
is not as protective as the bright.

Specifications for baking cadmium plated high strength steels to relieve hydrogen embrittlement tend to be vague, i.e., bake for 1 to 5 h at 149 to 204°C. In aerospace applications,
it is common practice to bake for 24 h at 191°C for the highest strength steels, or use a
sliding scale, depending on the strength level. It is important that the baking be performed
within 4 h of the plating operation. Although fractographic examination of the failed bolts
showed intergranular fracture origins (Zone 1) which occur as a result of HE and SCC, it is
unlikely that these mechanisms caused the failures. Both HE and SCC do not produce the
black oxide observed on the two failed bolts and the bolt found to contain a crack by MPI.
The only mechanism which could produce this oxide is thermal growth during heat treatment.
The stress durability test which was specifically designed to demonstrate effects of HE caused
by electroplating or exposure to other environments containing a source of hydrogen, showed
no failures after stressing at 80% of the UTS for 200 to 400 h. The preponderance of evidence
attributes the failure of the two bolts during installation to the presence of pre-existing quench
cracks.

Conclusions

1. Chemical analysis showed that the bolts were fabricated of AISI 8740 steel. The microstructure was tempered martensite typical of a quenched-and-tempered low alloy steel.
The material was clean with no major inclusions present. In the quenched-and-tempered
condition, the alloy should have a good combination of strength, toughness, and fatigue
resistance.

2. Radius measurements at the head-to-shank interface showed that approximately 50%
of the bolts were sharper than specified.

3. Of all the bolts examined from service and inventory, only one bolt failed magnetic
particle inspection due to the presence of a transverse crack near the head-to-shank radius.

4. The cadmium plating thickness on the two failed bolts varied. One bolt conformed to
the thickness and uniformity requirements of a Class 2 plating. The cadmium plate of the
other failed bolt was thinner and less uniform.

5. Mechanical property and hardness measurements of the two failed bolts and some from
inventory/service were within the range of specification requirements.

6. Torque tests comparable to those at the installation of bolts showed that all bolts tested
(eight from inventory and service) exceeded the minimum torque-to-failure requirement by
a factor of 2 to 3.

7. Stress durability tests for bolts which may have been exposed to hydrogen embrittlement conditions showed no bolt failures (29 bolts total) after preloading to 80% of the UTS for 200 to 400 h.

8. Light optical and scanning electron microscopy of fracture surfaces of the two failed bolts showed similar topographies consistent with the characteristics of quench cracks that existed prior to installation. The black oxide observed was attributed to the tempering treatment. (~649°C) after quenching.

9. Fractographic evidence in conjunction with stress durability test results favors a failure mode attributable to preexisting quench cracks and not HE or SCC.

Recommendations

1. Specify 100% magnetic particle inspections of bolts after the tempering operation.

2. Specify dull cadmium plate to mitigate the potential for delayed failures due to HE or SCC.

3. Specify a 24 h embrittlement relief baking at 191°C to insure removal/dispersal of hydrogen.

4. Alternatively, specify vacuum cadmium plate or ion-plating aluminum to eliminate the potential for hydrogen embrittlement.

5. Insure the radius at the head-to-shank interface conforms to specification requirements.

6. Review the vendors fabrication operation on-site with technical experts (metallurgist, chemist) from AMCOM and ARL.

7. Replace all existing bolts with new or reinspected inventory bolts to mitigate the possibility of undetected small quench cracks growing under firing loads.

References

[1] ASM Metals Handbook, Vol. 10, 8th ed., "Failure Analysis and Prevention," 1975, p. 74.
[2] ASM Metals Handbook, Vol. 9, 8th ed., "Fractography and Atlas of Fractographs," 1974, p. 308.

Fatigue and Fracture

Casey R. Brown[1] and Dale A. Wilson[1]

The Effect of Fasteners on the Fatigue Life of Fiber-Reinforced Composites

REFERENCE: Brown, C. R. and Wilson, D. A., **"The Effect of Fasteners on the Fatigue Life of Fiber-Reinforced Composites,"** *Structural Integrity of Fasteners: Second Volume, ASTM STP 1391.* P. M. Toor, Ed., American Society for Testing and Materials, West Conshohocken, PA, 2000, pp. 65–71.

ABSTRACT: The effects of stress concentrations on the fatigue life of fiber-reinforced graphite epoxy composites play a very important role in the aerospace industry. Primarily stress concentrations refer to the holes. However, fasteners are the primary manufacturing process for attachment of graphite epoxy panels in use in the aerospace industry today. These conditions of composites with and without fasteners greatly affect the fatigue properties of such panels.

In an attempt to understand the stress field created by fasteners in carbon-reinforced graphite composites, and how these stress fields affect the fatigue life of these composites, a fatigue life analysis program was conducted. It compared specimens with machined fastener holes and specimens with fasteners placed in them. In addition to the fatigue analysis performed on the specimens, photoelastic stress analysis is used to determine the stress field created from the use of fasteners. The difference in stress fields of specimens without fasteners compared to specimens with fasteners can be determined to allow designers to prolong the fatigue life of composites.

KEYWORDS: fasteners, fatigue composites, contact stresses, fastener fatigue failures

The use of fiber-reinforced graphite epoxy composite materials in the aerospace industry has increased significantly over the past 20 years. These materials are primarily used in the defense area of aerospace applications. These applications range from fighter jets such as the F/A-18 to attack helicopters. The primary use of composites on these aircraft is skin panels. However, these panels tend be attached to the metallic airframe with fasteners, such as rivets or bolts, introducing a significant stress concentration to the composite panels which in turn significantly affects their fatigue life.

This introduction of stress concentrations to the composite panel can also have different effects on the fatigue life of the panel depending on the type of cyclic loading present. While some applications present primarily tensile cyclic loading, others will introduce fully reversed tensile to compressive cyclic loading. The presence of holes and fasteners creates stress fields in the composite that alter depending upon the loading. Therefore, the stress ratio present has a very influential effect on the fatigue life of the composite. This is an area of technology that has not been extensively studied and its effects are not clearly known.

Experimental Program

A study has been initiated on the fatigue life and behavior of graphite epoxy laminates in the presence of stress concentrations at elevated temperatures. The program was developed

---

[1] Tennessee Technological University, Department of Mechanical Engineering, Cookeville, TN 38505.

to determine the *S-N* curves of two stress concentration conditions, in two different composite lay-ups, at a stress ratio of $R = -1$. The concentrations being studied are the presence of a mechanically-introduced circular hole, and the same hole filled with a fastener used for mechanical fastening to an aircraft airframe. The tests were conducted at 180°F. The two specimens tested are illustrated in Fig. 1. The two lay-ups were designated AML -33 and AML $+56$. The AML -33 specimens consisted of a lay-up percentagewise of 62/29/9 for the 0, 45, and 90° layers and are considered high strength composite lay-ups. The AML $+56$ specimens consisted of a lay-up percentagewise of 11/67/22 for the 0, 45, and 90° layers and are considered low strength composite lay-ups.

The fatigue tests were conducted using a MTS servo-hydraulic fatigue machine. In order to prevent buckling of the specimens during testing, a mechanical fixture was used with buckling guards. In order to maintain the constant 180°F (82°C) temperature, resistance heaters were attached to the mechanical fixture. The entire testing apparatus was enclosed to maintain moisture content of the specimens and to stabilize temperature.

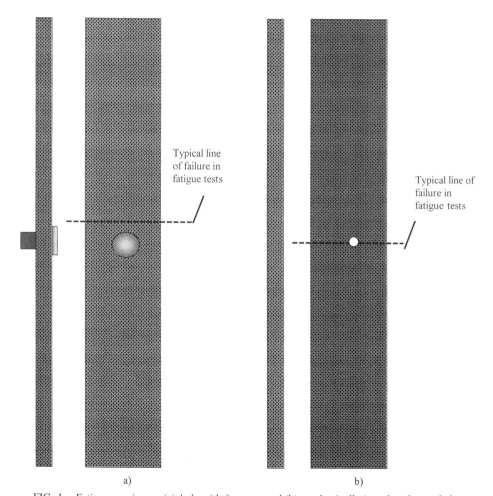

a)   b)

FIG. 1—*Fatigue specimens:* (a) *hole with fastener, and* (b) *mechanically-introduced open hole.*

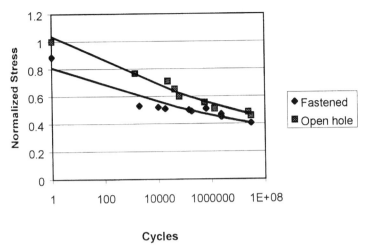

Cycles

FIG. 2—*Fatigue curve for AML −33.*

Results

An initial stress level of 60% of the ultimate compressive strength was chosen as a starting point to try to achieve failure of the specimens at 10 000 cycles. However, the cycles reached were quite different for the two stress concentrations. While the specimen with the fastener-filled hole had a higher nominal compressive strength in both lay-ups, at 60% of this value the rivet specimens for the two lay-ups failed at a much lower number of cycles than the specimens without rivets. For the next round of tests, 90% of the previous stress level for each specimen was used. The results of the fatigue tests are given in Figs. 2 and 3.

In all of the fastener specimens, the line of failure occurred at a point above or below the fastener itself, as shown in Fig. 1. The stresses have been calculated using the cross-sectional

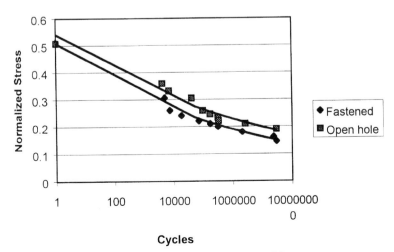

Cycles

FIG. 3—*Fatigue curve for AML +56.*

area along the line of failure. Therefore, the area of the hole is subtracted from the total cross-sectional area in the open hole specimens. The resulting stresses were then normalized. As shown in the figures, the separation in the curves increases as stress increases. The results indicate that the reduction in fatigue strength is load dependent.

Due to the location of the line of failure, the failure mode, and the load dependency, the hypothesis was drawn that the early failure of the fastener specimens was due to the presence of contact stresses between the composite material and the fastener. In the case of contact stresses, the maximum stress does not occur at the point of contact. The maximum stress occurs at a point away from the contact point at a certain radius distance. This stress concentration then acts as a point of nucleation for the fatigue crack to form. These types of fatigue cracks form in fiber-reinforced composites as delaminations; then the delaminations grow to the point where fiber failure occurs.

Mathematically, this problem can be modeled simply as a cylinder inside a cylindrical seat [1]. Using this mathematical model requires specific assumptions to be made, such as the material being homogeneous and isotropic along with others. As seen in Fig. 4, the contact stresses occur due to contact between the composite material and the fastener. In this case, the contact area is a narrow rectangle of width 2b and length L, which is the thickness of the composite material. The maximum contact stress occurs at a distance b from the point of contact.

By altering certain parameters of the physical design, it is possible to greatly reduce the resulting contact stresses that are present. If the composite material properties, thickness, and loading conditions are fixed quantities that cannot be altered in the design, it is possible to alter the fastener material and the fastener clearance. The resulting reductions in contact stresses using different fastener materials and fastener clearances are shown in Figs. 5 and 6. Figure 5 uses a steel fastener and alters the clearance to examine the effect. Figure 6 uses a 0.001 clearance and alters the fastener material to examine the effect. Due to the high loading conditions for the AML −33, an equivalent specimen at its highest fatigue load with a steel fastener and one thousandths clearance would result in a contact stress approximately 1.4 times the nominal stress. Under these conditions, the total stress is more than doubled

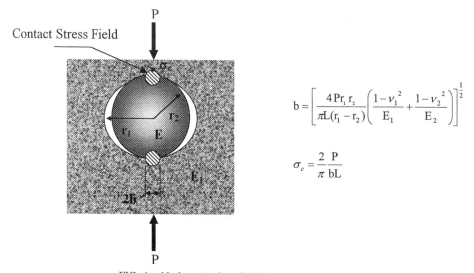

$$b = \left[\frac{4 P r_1 r_2}{\pi L (r_1 - r_2)} \left(\frac{1 - \nu_1^2}{E_1} + \frac{1 - \nu_2^2}{E_2} \right) \right]^{\frac{1}{2}}$$

$$\sigma_c = \frac{2}{\pi} \frac{P}{bL}$$

FIG. 4—*Mathematical model of contact stresses [1].*

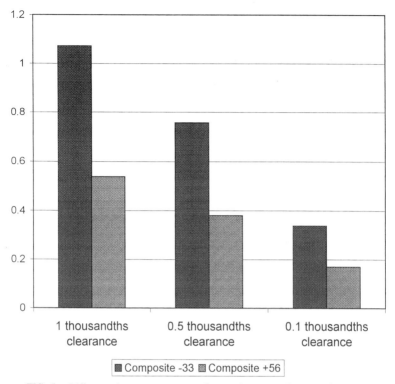

FIG. 5—*Difference in contact stresses due to changes in fastener clearance.*

by the presence of the contact stresses. However, by altering the design parameters, the magnitude of the resulting contact stress can be greatly reduced. The most ideal conditions, such as using an aluminum fastener with a 0.1 thousandths clearance, result in a contact stress that is approximately ⅓ of the normal stress. Therefore, the total stress would only be slightly increased due to the contact stresses. As shown by the figures, the fastener clearance has a much more profound effect on the contact stresses than the fastener material. However, the fastener material does have a significant effect.

The results of the mathematical model gave the point of highest stress that very closely matched the failure areas for the fastener specimen. However, that did not conclusively prove that the contact stresses were present. Photoelastic testing was performed to determine the stress field present in the case of an open hole versus a hole with a steel rivet. Since the photoelastic coating could not be used with the composite specimens because of the buckling guides, the testing was performed using an aluminum specimen in compression. However, this testing was inconclusive in proving the presence of contact stresses. Due to problems with buckling, loads were not significant enough to cause detectable amounts of contact stresses. However, literature research found that classical photoelastic testing of a tension member with a rivet placed in a hole had been performed by Coker and Filan [2]. Their research found that in the case of tension, a rivet caused significant amounts of contact stress in the tension member. Therefore, it can be assumed that this also occurs in compression.

While the presence of the fastener has been shown to be detrimental in the case of low cycle fatigue, its presence may be beneficial at high cycle fatigue. As can be seen from the

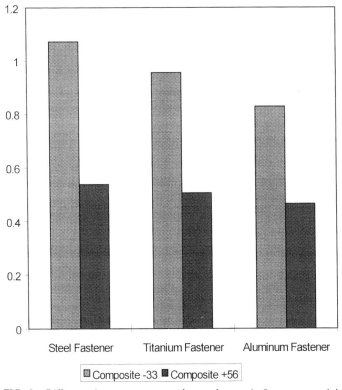

FIG. 6—*Difference in contact stresses due to changes in fastener material.*

mathematical model, as axial loading decreases so too do the contact stresses. The failure mode for the specimens with the circular hole has been primarily due to a fatigue crack nucleating as a delamination. The presence of the rivet tends to constrain delamination. At what point the contact stresses reduce enough for the rivet to increase fatigue life is unknown, if at all.

Conclusions

As fiber-reinforced graphite epoxy composite laminates become increasingly important in the aerospace industry, prediction of fatigue life in the presence of stress concentration factors such as fasteners becomes crucial. As shown by earlier work and by experiments currently being conducted, the failure mode of these composites is very complex. It has also been shown that it may be possible to improve the fatigue life of composites. At the present time, the results and observations merely indicate possible causes for failure, and areas where improvements in fatigue could eventually be made. Much work needs to be done to create an accurate model for use in composite materials. However, the work here is intended to give designers the opportunity to increase component fatigue life by significant reduction of the contact stresses that are present.

References

[*1*] Ugural, A. C. and Fenster, S. K., *Advanced Strength and Applied Elasticity,* 3rd ed., Prentice Hall PTR, Englewood Cliffs, NJ, 1995.
[2] Coker, E. G. and Filon, L. N. G., *Treatise on Photo-Elasticity,* 2nd ed., Cambridge University Press, U.K., 1957.

David F. Alexander,[1] George W. Skochko,[1] Wayne R. Andrews,[2] and R. Scott Briody[2]

Fatigue Testing of Low-Alloy Steel Fasteners Subjected to Simultaneous Bending and Axial Loads

REFERENCE: Alexander, D. F., Skochko, G. W., Andrews, W. R., and Briody, R. S., "**Fatigue Testing of Low-Alloy Steel Fasteners Subjected to Simultaneous Bending and Axial Loads,**" *Structural Integrity of Fasteners: Second Volume, ASTM STP 1391,* P. M. Toor, Ed., American Society for Testing and Materials, West Conshohocken, PA, 2000, pp. 72–84.

ABSTRACT: Fatigue testing of high strength, low-alloy steel fasteners has been completed with simultaneous unidirectional bending and axial loads that demonstrates a potentially beneficial effect on fatigue life by accounting for bending. Fatigue tests were performed in load control on fasteners with failures in rolled and cut/machined threads. The objective of the testing was to determine if the fatigue strength of steel fasteners subjected to high bending and axial stresses is higher than in conventional axial, load controlled tests. Test results in the range of 10^4 to 10^6 cycles indicate that the fatigue lives of fasteners with a 2:1 ratio of bending to axial stress is from two to ten times as great as that of fasteners subjected to axial loading only. A special fixture design was used for the testing. The fixture permits an adjustable offset between the fastener and clevis centerlines, thus producing an eccentric loading. The eccentric load results in simultaneous application of bending and axial loads. The magnitude of the bending and axial loads was verified by strain gage measurements on the reduced shank diameter of the test specimens. Fatigue life was determined based on failure cycles (specimen separation). Stresses corresponding to failure cycles were based on the sum of the alternating axial and bending stresses at the half-life of the test specimen. A low degree of data scatter and good agreement with published data were observed for the axial load tests performed with the fixture set at zero offset between the clevis and fastener centerlines.

KEYWORDS: fatigue testing, low-alloy steel, fasteners, simultaneous bending and axial loads

During service, fasteners may be subjected to combinations of axial and bending cyclic loads. Fatigue evaluation of fasteners based on axial load fatigue data is judged to be conservative when evaluating fasteners loaded by significant cyclic bending loads. To determine the difference between the combined axial-bending and axial fatigue strength only of low-alloy steel fasteners, a limited number of fatigue tests at 70°F (21°C) were conducted to failure (complete separation) in the threads. A total of eighteen ³⁄₄-in. 10 UNRC-2A bolts were fatigue tested to failure; nine bolts with conventionally machined threads (four axial fatigue and five axial-bending fatigue), and nine bolts with rolled threads (five axial fatigue and four axial-bending fatigue) were fatigue tested. The ratio of the alternating bending stress

[1] Engineering manager and senior engineer, respectively, Bechtel Plant Machinery, Inc., P.O. Box 1021, Schenectady, NY 12301.
[2] Senior engineer and senior specialist, respectively, Materials Characterization Laboratory (MCL), 704 Corporations Park, Scotia, NY 12302.

TABLE 1—*Chemistry of fatigue tested low-alloy steel materials.*

Material	AISI 4340	AISI 4140
C	0.38–0.43	0.396
Mn	0.65–0.85	0.400
P	0.015 max	0.013
S	0.010 max	0.013
Si	0.020–0.035	0.240
Ni	1.65–2.00	0.040
Cr	0.70–0.90	1.060
Mo	0.20–0.30	0.014

to the alternating axial stress at maximum cyclic loads was set equal to 2:1. The bending fatigue tests are compared to axial fatigue tests that were previously evaluated [1].

Comparison of the unnotched bending and axial fatigue strengths of low-alloy steel bars showed that depending on the chemistry, type, and strength of steel, the bending fatigue endurance of such steels could be up to 1.5 times as high as their axial fatigue strength [2,3]. A further evaluation of simultaneous bending and torsional fatigue strength of AISI 4340 steel round bars with circumferential U-notches was performed [4]. No comparisons are known between the axial and simultaneous bending and axial fatigue strengths of threaded fasteners such as bolts or studs, with prototypically loaded threads.

Fastener Materials

The materials of the tested fasteners were chosen to be low-alloy steels since existing data evaluated 198 axial fatigue tests of low-alloy steel fasteners ranging in size from 1 to 5¹⁄₁₆ in. (2.54 to 12.85 cm) in diameter [1]. In this test program, the machined thread fasteners were made of AISI 4340 low-alloy steel, and the rolled threaded fasteners were made of AISI 4140 steel. The chemistries of the fastener materials and their tensile properties are shown in Tables 1 and 2, respectively.

The rolled thread fasteners were heat treated at 1050°F (566°C) in argon for 4½ h, and air cooled to room temperature to reduce the ultimate tensile strength to be comparable to that of the machined thread specimens, and to reduce the compressive residual stresses at the thread roots due to thread rolling. Rockwell C hardness readings were made on the bolt

TABLE 2—*Tensile properties of tested fastener materials.*

Property	AISI 4340	AISI 4140
Ultimate tensile strength, ksi*	153–164	142 (See Note 1)
Yield strength, ksi*	137–148	⋯
% Elongation	17.5–19.5	⋯
% Reduction in area	59.9–62.2	⋯

* 1 ksi = 6.895 MPa.

NOTE 1—The ultimate tensile strength shown for the AISI 4140 material is based on Rockwell C (R_c) hardness measurements obtained on top of bolt hex head after heat treatment. Five R_c measurements from each test specimen were used to determine a range of values between $29.7 \leq R_c \leq 30.6$, resulting in a corresponding ultimate tensile strength of $141 \text{ ksi} \leq S_u \leq 145 \text{ ksi}$. A tensile test of one specimen indicated an ultimate strength of 138 ksi.

head hex flats of all rolled thread fasteners to confirm their ultimate tensile strengths. They were 138 to 145 ksi (952 to 1000 MPa). Three different heats of machined thread fastener material were heat treated and tempered at 1000°F (538°C) minimum for one hour prior to machining the threads with resultant ultimate tensile strengths of 153.0 to 164.0 ksi (1055 to 1131 MPa).

Test Specimens

All test specimens used for the new fatigue tests were ¾ in.-10 UNRC-2A bolts modified by machining a reduced shank as shown in Fig. 1. Rolled thread test specimens of AISI 4140 material were made from four inch long "off the shelf" bolts trimmed and machined as shown in Fig. 1. Machined/cut thread specimens were machined from AISI 4340 bar stock. All bolts were serially marked for traceability after final machining. The large fillet radii and 2.25 in. (5.71 cm) long, straight shank of the bolts shown in Fig. 1 allowed the placement of strain gages at two locations along their lengths, away from any stress concentrations. Strain gages at the two elevations were used to measure the variation of bending strains in the shank, which were then used to compute stresses in the first loaded thread assuming a linear variation of strains along the bolt length and linearity between stress and strain.

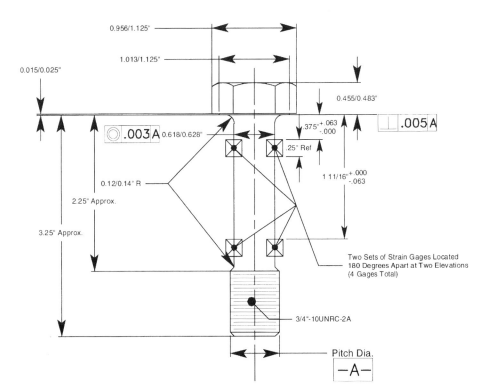

FIG. 1—*Bolt geometry and strain gage locations. (Note: 1 in. = 2.54 cm.)*

Loosening of the bolt was prevented by a hex recess machined into the fixture at the top of the bolt, and a preload locking plate was used to restrain the nut at the bottom of the specimen.

Test Fixture

The test fixture, shown in Fig. 2, simultaneously applies an axial load P and a bending moment M to the bolt. The bending moment was produced by an offset between the bolt and clevis centerlines resulting in an eccentric load, bending and stretching the bolt, as shown in Fig. 2. The offset between the bolt and clevis centerlines allowed offset adjustments of up to ½ in. (12.7 mm), and was set by a vernier scale attached to the fixture. The adjustable offset was achieved within a sliding dovetail joint between the clevis and the adapter plates. Capabilities of the fixture allowed different offsets between the bolt centerline and the upper

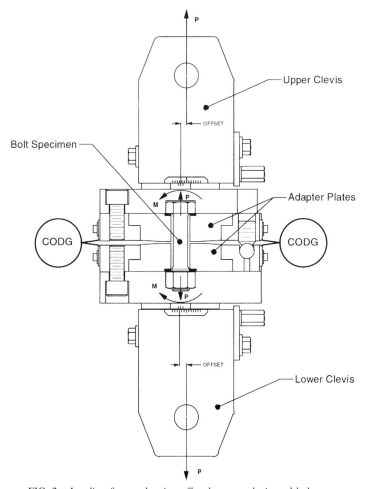

FIG. 2—*Loading fixture showing offset between clevis and bolt axes.*

or lower clevis, allowing a bending gradient along the bolt length. All bending tests in this program were conducted with the same offset for the upper and lower clevis.

The nominal stress applied to the bolt by the fixture was the sum of the axial and bending stresses given by:

$$\sigma = \sigma_{membrane} + \sigma_{bending} = P/A + Mc/I \tag{1}$$

The above equation is a function of the axial load, P, the bolt diameter, D, and the offset or eccentricity, e, and reduces to the following:

$$\sigma = 4P/\pi D^2 (1 + 8e/D) \tag{2}$$

Fatigue Tests

Axial fatigue tests were performed on low-alloy steel fasteners for axial loading to benchmark the fixture and test specimens against existing fastener fatigue test data. The tests were repeated for simultaneous bending and axial loads. This allowed a comparison of the results of the axial fatigue tests and the simultaneous bending and axial fatigue tests. Each test was set up, and the test cycles were applied until fracture of the specimen. Fasteners were preloaded to ensure no unloading during cyclic testing.

Instrumentation and Test Setup

Instrumentation used for the fatigue tests consisted of strain gages on the bolt shanks, Crack opening displacement gages (CODGs) on the outside of the fixture, and the load cell. Strain gages were used to measure the magnitude of axial and bending strains on the bolt shank. CODGs were used to minimize bending during the set up of axial tests. The load cell was used to verify the magnitude of axial strains.

The setup for each test consisted of installation of an instrumented bolt into the fixture, and verification of the desired test parameters through the application of initial (5 to 10) manual load/unload cycles. The use of a graphical data reduction interface (LabView [5] in this case) facilitated the test setup verification by providing a summary of all pertinent test data as shown in Fig. 3 and also had the capability to transfer test data (measured during set up) to a spreadsheet for test documentation.

The control parameter used for each test was the strain range. Existing fastener fatigue data [1] were used as a basis to establish fatigue test strain ranges that produced failures between 1000 and 1 000 000 cycles. Once the strain range was established for a particular test, minimum and maximum stresses and strains were determined as shown in Appendix 1. Axial loads based on the shank cross sectional area were then computed, and used as a starting point to begin the application of manual load and unload cycles. All fatigue tests were performed with a stress ratio of 0.1 (i.e., the ratio of minimum to maximum stress). This ensured that the bolts were never unloaded.

During the setup of simultaneous bending and axial load tests, the ranges of bending and axial strains were monitored to ensure that bending strain range was approximately twice the range of axial strains. This was referred to as the "bending to membrane ratio." This ratio was controlled through the adjustable offset between the centerlines of the clevis and the fastener. The offset was adjusted until the desired strain range and bending to membrane ratio was obtained. The offset had to be established during the test setup since it could not be adjusted during the cyclic portion of the tests. All simultaneous bending and axial load tests were performed with a bending to membrane ratio of approximately 2:1.

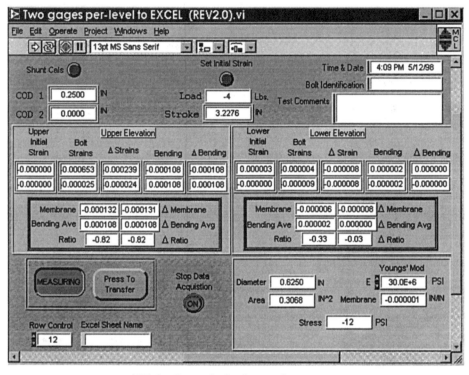

FIG. 3—*Screen display for sample test setup.*

Test Execution

Once the test setup was completed and the test parameters verified, a Wavetek waveform generator was used to control the cyclic portion of the test, along with the test machine controls. A sinusoidal waveform was applied at a rate of between 3 and 5 Hz. Each test began with the fastener at the mean load, which was adjusted within the first few hundred cycles until the desired strain range was observed. Strains were measured at a rate of 10 to 16 times per cycle and recorded every 100 cycles in an ASCII file for subsequent retrieval and analysis. Although offsets were not adjusted during the simultaneous bending and axial load tests, the strain range was continuously monitored during the early test cycles and small adjustments were made to ensure consistency of strain range. Fatigue tests were terminated immediately after fracture of the test specimens.

Test Results

Fatigue test results of low-alloy steel bolts are tabulated in Table 3 and plotted in Figs. 4 and 5 for rolled and cut/machined threads, respectively. It is noted that the new data are limited in number and are intended only to indicate qualitatively the effect of bending and not to quantify this effect. New data represent recent tests that were performed at the Materials Characterization Laboratory.

The results of axial fatigue tests of bolts with rolled threads in Fig. 4 show good correlation with existing axial thread fatigue data, as indicated by the proximity of the New Data/Rolled

TABLE 3—*Fatigue test data summary.*

Test	Temp	Thread Type	Fastener Size	Failed at	Axial Stress ksi**	Bending Stress, ksi**	S_{range} ksi**	S_{alt} ksi**	S_u ksi**	S_{alt}/S_u	N_{fail} Cycles
R_Ax_1	70	Rolled	3/4-10UNRC	Thread	63.0	0.0	63.0	31.5	142	0.222	8,552
R_Ax_2	70	Rolled	3/4-10UNRC	Thread	20.0	2.6	22.5	11.3	142	0.079	735,000
R_Ax_3	70	Rolled	3/4-10UNRC	Thread	49.3	2.0	51.3	25.7	142	0.181	45,000
R_Ax_4	70	Rolled	3/4-10UNRC	Thread	39.7	0.9	40.6	20.3	142	0.143	101,086
R_Bnd_1	70	Rolled	3/4-10UNRC	Thread	15.2	28.8	44.0	22.0	142	0.155	875,000
R_Bnd_2	70	Rolled	3/4-10UNRC	Thread	20.9	37.8	58.7	29.4	142	0.207	85,000
R_Bnd_3	70	Rolled	3/4-10UNRC	Thread	24.4	44.1	68.5	34.2	142	0.241	40,900
R_Bnd_4	70	Rolled	3/4-10UNRC	Thread	30.7	52.9	83.6	41.8	142	0.294	24,900
R_Bnd_5	70	Rolled	3/4-10UNRC	Thread	30.4	56.5	86.9	43.5	142	0.306	19,000
C_Ax_1	70	Machined	3/4-10UNRC	Thread	34.9	N/A*	34.9	17.5	159	0.110	140,116
C_Ax_2	70	Machined	3/4-10UNRC	Thread	56.8	N/A*	56.8	28.4	159	0.178	49,700
C_Ax_3	70	Machined	3/4-10UNRC	Thread	24.4	N/A*	24.4	12.2	155	0.079	609,500
C_Ax_4	70	Machined	3/4-10UNRC	Thread	83.9	N/A*	83.9	42.0	155	0.271	8,300
C_Ax_5	70	Machined	3/4-10UNRC	Thread	64.1	N/A*	64.1	32.1	155	0.207	12,793
C_Bnd_1	70	Machined	3/4-10UNRC	Thread	32.5	61.1	93.6	46.8	163	0.287	17,477
C_Bnd_2	70	Machined	3/4-10UNRC	Thread	18.6	39.3	57.9	29.0	163	0.178	127,008
C_Bnd_3	70	Machined	3/4-10UNRC	Thread	14.9	28.0	42.9	21.5	154	0.139	430,000
C_Bnd_4	70	Machined	3/4-10UNRC	Thread	24.0	41.4	65.4	32.7	154	0.212	69,700

* Not Measured

**1 ksi = 6.895 MPa

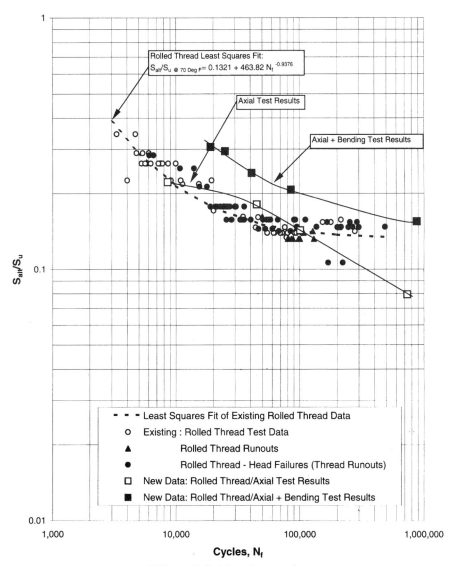

FIG. 4—*Rolled thread test results.*

Thread/Axial Test Results (shown as open boxes) in comparison to the existing data (shown as open circles). It is noted that all the new axial fatigue tests results were obtained using the fixture shown in Fig. 2 and setting the offset to zero.

Combined bending and axial load fatigue test results of bolts with rolled threads are shown in Fig. 4 as black boxes. These data show consistently improved fatigue life compared to axial fatigue tests. The improved fatigue strength of bolts tested with combined bending and axial stresses range roughly from a factor of 4 to 20 on cycles over the range of 10^4 to 10^6 cycles.

Figure 5 shows the new axial fatigue tests performed on bolts with cut/machined threads (open boxes), with failures ranging between 8000 and 600 000 cycles. These showed rea-

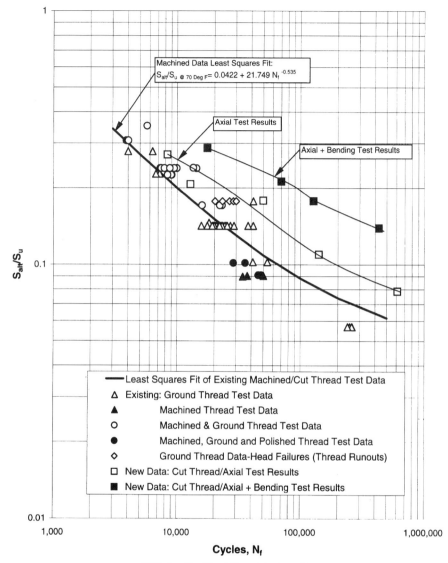

FIG. 5—*Machined thread test results.*

sonable correlation with available axial thread data below 40 000 cycles but fall above the available data at 600 000 cycles. Simultaneous bending and axial fatigue tests, shown as black boxes, show improved fatigue life compared to both the existing axial bolt thread fatigue data as well as to the new axial bolt fatigue tests. The improvement appears to be greatest at small alternating stresses and high cycles to failure. Tests with simultaneous bending and axial fatigue strains showed improvement ranging roughly from 3 to 5 on cycles for fatigue lives of 10^4 to 2×10^5 cycles, consistent with the results of the rolled thread data.

It is noted that the new data were too limited to justify the development of a fatigue curve for combined bending and axial loading. However, if enough data were available, a "case basis" fatigue curve could be developed for the explicit bending to membrane ratio tested. This curve could then be used to quantitatively assess the fatigue benefit of using the "case basis" combined loading curve in relation to the typical fatigue curve based on axial loading alone.

Conclusions

Correlation with Existing Axial Test Data

New axial fatigue test results of the rolled and of the machined threads appear to correlate well with the existing axial fatigue data of both rolled, and of ground and machined threads of low-alloy steels. At intermediate (10^4 to 10^5) cycles to failure, the new axial data for rolled threads correlate well with the existing data (Fig. 4); however, a single new data point at 700 000 cycles falls below the existing rolled thread data. A comparison of the machined axial data at 8000 to 40 000 cycles to failure indicates that the new data correlate well with existing data (Fig. 5). Two new tests with failures at 150 000 and at 600 000 cycles, respectively, fall considerably above the existing test results, but existing data are sparse in this high cycle regime.

Correlation of Axial with Axial Plus Bending Fatigue Data

For a bending to membrane ratio of 2:1, there is a significant improvement in fatigue life when axial plus bending test results are compared to purely axial data. Although there was no intent to quantify the improvement, given the limited data, it appears that the improvement is greater for rolled threads (4 to 20 on cycles) than for machined (3 to 5 on cycles) in the range of 10^4 and 10^6. Fatigue strength improvement appears more pronounced at higher cycles than at the lower cycles to failure. Axial and axial plus bending results appear to converge below 10^4 cycles, but data are too limited to warrant extrapolating a curve fit.

Limitations of Test Results and Potential Applications

As previously noted, the new test results discussed above are based on limited data (18 total specimens of a single size). However, the authors judge that the objective of the new testing was accomplished, i.e., to demonstrate, but not to quantify, that a potential benefit on fatigue strength can be realized by accounting for bending stresses when they are a significant part of the total stress range. Any quantitative applications must rely on a more rigorous testing effort focused on alternating stress states that are representative of expected parameters.

References

[1] Skochko, G. W. and Herrmann, T. P., "Review of Factors that Affect Fatigue Strength of Low-Alloy Fasteners," *ASTM STP 1236 Structural Integrity of Fasteners,* P. M. Toor, Ed., American Society for Testing and Materials, West Conshohocken, PA, 1995, pp. 32–47.
[2] Heywood, R. B., *Designing Against Fatigue of Metals,* Reinhold Publishing Corp., New York, 1962 (Sections 2.6 and 2.7).
[3] Banantine, J. A., Comer, J. J., and Handrock, J. L., *Fundamentals of Metal Fatigue Analysis,* Prentice Hall, 1990, p. 13.
[4] Kececioglu, D., Chester, L. B., and Dodge, T. M., "Combined Bending-Torsion Fatigue Reliability of AISI 4340 Steel Shafting with Kt = 2.34," *Journal of Engineering for Industry, Trans. ASME,* May 1975, pp. 748–760.

[5] LabVIEW® for Windows, National Instruments Corporation, 6504 Bridge Point Parkway, Austin, TX 78730-5039.

Appendix 1—Test setup calculations.

EQUATIONS AND DEFINITIONS:

S_u = Ultimate Stress, psi (Mpa) S_{max} = Maximum Stress, psi (Mpa)   R = Stress Ratio

S_{alt} = Alternating Stress, psi (Mpa) S_{min} = Minimum Stress, psi (Mpa)

$$\text{Ratio} := \frac{S_{alt}}{S_u}$$

$$S_{alt} := \frac{(\sigma_{max} - \sigma_{min})}{2}$$

$$R := \frac{\sigma_{min}}{\sigma_{max}}$$

$$\text{Range} := \sigma_{max} - \sigma_{min}$$

$$\text{Range} := \sigma_{max} \cdot (1 - R)$$

$R := .1$

$S_u := 140000$ psi (965.5 MPa)

$\text{Ratio}_1 := .27$

$\text{Ratio}_2 := .1$

$\text{Ratio}_3 := .06$

$S_{alt\_1} := \text{Ratio}_1 \cdot S_u$

$S_{alt\_2} := \text{Ratio}_2 \cdot S_u$

$S_{alt\_3} := \text{Ratio}_3 \cdot S_u$

$S_{alt\_1} = 37800$ psi (260.1 MPa)

$S_{alt\_2} = 14000$ psi (96.6 MPa)

$S_{alt\_3} = 8400$ psi (57.9 MPa)

$S_{Range\_1} := 2 \cdot S_{alt\_1}$

$S_{Range\_2} := 2 \cdot S_{alt\_2}$

$S_{Range\_3} := 2 \cdot S_{alt\_3}$

$S_{Range\_1} = 75600$ psi (521.4 MPa)

$S_{Range\_2} = 28000$ psi (193.1 MPa)

$S_{Range\_3} = 16800$ psi (115.9 MPa)

$$\sigma_{max\_1} := \frac{S_{Range\_1}}{(1 - R)}$$

$$\sigma_{max\_2} := \frac{S_{Range\_2}}{(1 - R)}$$

$$\sigma_{max\_3} := \frac{S_{Range\_3}}{(1 - R)}$$

$\sigma_{max\_1} = 84000$ psi (579.3 MPa)

$\sigma_{max\_2} = 31111.1$ psi (214.6 MPa)

$\sigma_{max\_3} = 18666.7$ psi (128.1 MPa)

$\sigma_{min\_1} := R \cdot \sigma_{max\_1}$

$\sigma_{min\_2} := R \cdot \sigma_{max\_2}$

$\sigma_{min\_3} := R \cdot \sigma_{max\_3}$

$\sigma_{min\_1} = 8400$ psi (57.9 MPa)

$\sigma_{min\_2} = 3111.1$ psi (21.1 MPa)

$\sigma_{min\_3} = 1866.7$ psi (12.9 MPa)

Sample Axial Load Calculation For 3/4-10UNC 2A Threads:

$n := 10$ Threads per Inch   (0.394 Threads per mm)

$E := .685$ Basic Pitch Diameter, in   (17.40 mm)

$$A_s := .7854 \cdot \left(E - \frac{0.32476}{n}\right)^2$$ Effective Tensile Stress Area, in^2

$A_s = 0.334$ in^2 (212.9 mm^2)

$\text{Load}_{max\_1} := \sigma_{max\_1} \cdot A_s$

$\text{Load}_{max\_2} := \sigma_{max\_2} \cdot A_s$

$\text{Load}_{max\_3} := \sigma_{max\_3} \cdot A_s$

$\text{Load}_{max\_1} = 28090.7$ lbf

(6315 N)

$\text{Load}_{max\_2} = 10404$ lbf

(2339 N)

$\text{Load}_{max\_3} = 6242.4$ lbf

(1413 N)

$\text{Load}_{min\_1} := R \cdot \text{Load}_{max\_1}$

$\text{Load}_{min\_2} := R \cdot \text{Load}_{max\_2}$

$\text{Load}_{min\_3} := R \cdot \text{Load}_{max\_3}$

$\text{Load}_{min\_1} = 2809.1$ lbf

(613.5 N)

$\text{Load}_{min\_2} = 1040.4$ lbf

(234 N)

$\text{Load}_{min\_3} = 624.2$ lbf

(141.3 N)

Appendix 1—(*Continued*).

$E_{Mod} := 29900000$ psi $D_{Shank} := 0.623$ in $A_{Shank} := \pi \cdot \dfrac{D_{Shank}^2}{4}$ $A_{Shank} = 0.305$ in^2

(206,207 MPa)   (15.82 mm)   (196.8 mm^2)

$Load_{max\_1} = 28090.7$ lbf $\varepsilon_{max\_1} := \dfrac{Load_{max\_1}}{A_{Shank} \cdot E_{Mod}}$ $\varepsilon_{min\_1} := R \cdot \varepsilon_{max\_1}$

(6315 N)

$\varepsilon_{max\_1} = 0.003082$ $\varepsilon_{min\_1} = 0.000308$ $\varepsilon_{range\_1} := \varepsilon_{max\_1} - \varepsilon_{min\_1}$ $\varepsilon_{range\_1} = 0.002774$

$Load_{max\_2} = 10404$ lbf $\varepsilon_{max\_2} := \dfrac{Load_{max\_2}}{A_{Shank} \cdot E_{Mod}}$ $\varepsilon_{min\_2} := R \cdot \varepsilon_{max\_2}$

(2339 N)

$\varepsilon_{max\_2} = 0.001141$ $\varepsilon_{min\_2} = 0.000114$ $\varepsilon_{range\_2} := \varepsilon_{max\_2} - \varepsilon_{min\_2}$ $\varepsilon_{range\_2} = 0.001027$

$Load_{max\_3} = 6242.4$ lbf $\varepsilon_{max\_3} := \dfrac{Load_{max\_3}}{A_{Shank} \cdot E_{Mod}}$ $\varepsilon_{min\_3} := R \cdot \varepsilon_{max\_3}$

(1413 N)

$\varepsilon_{max\_3} = 0.000685$ $\varepsilon_{min\_3} = 0.000068$ $\varepsilon_{range\_3} := \varepsilon_{max\_3} - \varepsilon_{min\_3}$ $\varepsilon_{range\_3} = 0.000616$

Sample Bending Load Calculation

$Ratio_{B1} := .3$ $R = 0.1$

$S_{Alt\_B1} := Ratio_{B1} \cdot S_u$ $S_{Alt\_B1} = 42000$ psi   (289.6 MPa)

$S_{Range\_B1} := 2 \cdot S_{Alt\_B1}$ $S_{Range\_B1} = 84000$ psi   (579.3 MPa)

$\sigma_{max\_B1} := \dfrac{S_{Range\_B1}}{(1 - R)}$ $\sigma_{max\_B1} = 93333.3$ psi   (643.7 MPa)

$\sigma_{min\_B1} := R \cdot \sigma_{max\_B1}$ $\sigma_{min\_B1} = 9333.3$ psi $\sigma_{mem\_B1} := \dfrac{\sigma_{max\_B1}}{3}$

(64.4 MPa)

$A_s = 0.334$ in^2   (215.5 mm^2) $\sigma_{bnd\_B1} := \sigma_{mem\_B1} \cdot 2$

$D_s := \sqrt{4 \cdot \dfrac{A_s}{\pi}}$ $D_s = 0.653$ in   (16.59 mm)

$Sect_{mod\_s} := \pi \cdot \dfrac{D_s^3}{32}$ $Sect_{mod\_s} = 0.027$ in^3 $M := \sigma_{bnd\_B1} \cdot Sect_{mod\_s}$
                (442.5 mm^3)

$D_{Shank} = 0.623$ in   (15.8 mm) $A_{Shank} = 0.305$ in^2 $M = 1697.2$ in*lbf
                (196.8 mm^2)   (191.8 N*m)

$Sect_{mod\_Shank} := \pi \cdot \dfrac{D_{Shank}^3}{32}$ $Sect_{mod\_Shank} = 0.024$ in^3   (393.3 mm^3)

Appendix 1—(*Continued*).

For 2:1 Bending, the Membrane stess is 1/3 of the total, therefore:

Load Calculation

$$P_{B1} := A_s \cdot \frac{\sigma_{max\_B1}}{3} \qquad P_{B1} = 10404 \quad \text{lbf} \qquad P_{B1\_min} := .1 \cdot P_{B1} \qquad P_{B1\_min} = 1040.4 \quad \text{lbf}$$
$$\text{(2339 N)} \qquad\qquad\qquad\qquad\qquad\qquad \text{(233.9 N)}$$

Shank Strain Calculation

$$\sigma_{mem\_sh} := \frac{P_{B1}}{A_{Shank}} \qquad \sigma_{mem\_sh} = 34129.8 \ \text{psi} \qquad \varepsilon_{mem\_sh} := \frac{\sigma_{mem\_sh}}{E_{Mod}} \qquad \varepsilon_{mem\_sh} = 0.001141$$
$$\text{(235.4 MPa)}$$

$$SM_{Shank} := \pi \cdot \frac{D_{Shank}^3}{32} \qquad SM_{Shank} = 0.024 \ \text{in}^3$$
$$\text{(393.3 mm}^3\text{)}$$

$$\sigma_{bnd\_sh} := \frac{M}{SM_{Shank}} \qquad \sigma_{bnd\_sh} = 71494.5 \ \text{psi} \qquad \varepsilon_{bnd\_sh} := \frac{\sigma_{bnd\_sh}}{E_{Mod}} \qquad \varepsilon_{bnd\_sh} = 0.002391$$
$$\text{(453.1 MPa)}$$

D. M. Oster[1] and W. J. Mills[1]

Stress Intensity Factor Solutions for Cracks in Threaded Fasteners*

REFERENCE: Oster, D. M. and Mills, W. J., "**Stress Intensity Factor Solutions for Cracks in Threaded Fasteners,**" *Structural Integrity of Fasteners: Second Volume, ASTM STP 1391*, P. M. Toor, Ed., American Society for Testing and Materials, West Conshohocken, PA, 2000, pp. 85–101.

ABSTRACT: Nondimensional stress intensity factor (*K*) solutions for continuous circumferential cracks in threaded fasteners were calculated using finite-element methods that determined the energy release rate during virtual crack extension. Assumed loading conditions included both remote tension and nut loading, whereby the effect of applying the load to the thread flanks was considered. In addition, *K* solutions were developed for axisymmetric surface cracks in notched and smooth round bars. Results showed that the stress concentration of a thread causes a considerable increase in *K* for shallow cracks, but has much less effect for longer cracks. In the latter case, values of *K* can be accurately estimated from *K* solutions for axisymmetric cracks in smooth round bars. Nut loading increased *K* by about 50% for shallow cracks, but this effect became negligible at crack depth-to-minor diameter ratios (*a*/*d*) greater than 0.2. An evaluation of thread root acuity effects showed that the root radius has no effect on *K* when the crack depth exceeds 2% of the minor diameter. Closed-form *K* solutions were developed for both remote-loading and nut-loading conditions, and for a wide range of thread root radii. The *K* solutions obtained in this study were compared with available literature solutions for threaded fasteners as well as notched and smooth round bars.

KEYWORDS: threaded fasteners, stress intensity factor solutions, axisymmetric cracks, axisymmetric finite-element analysis

Nomenclature

a	Crack depth, in. (mm)[2]
a'	Assumed crack depth in *K* estimation procedure, in. (mm)
Δa	Increment of crack extension, in. (mm)
d	Minor diameter of thread or notch, in. (mm)
D	Major diameter of thread or outer diameter of notched or smooth bar, in. (mm)
E	Elastic modulus, psi (MPa)
F	Nondimensional stress intensity factor, $F = K/(\sigma\sqrt{\pi a})$

[1] Bettis Atomic Power Laboratory, West Mifflin, PA 15122.
* This report was prepared as an account of work sponsored by the United States Government. Neither the United States, nor the United States Department of Energy, nor any of their employees, nor any of their contractors, subcontractors, or their employees, makes any warranty, express or implied, or assumes any legal liability or responsibility for the accuracy, completeness or usefulness of any information, apparatus, product or process disclosed, or represents that its use would not infringe privately owned rights.
[2] Dimensions for the unified standard threads studied in this paper are defined in inches. To be consistent with these dimensions, English units are used as the primary measure. Where appropriate, the metric equivalent has been included in parentheses.

F_ρ Thread root radius correction factor
G Strain energy release rate, in.-lb/in.^{2} (kJ/m^2)
K Stress intensity factor, psi$\sqrt{\text{in.}}$ (MPa$\sqrt{\text{m}}$)
K_t Stress concentration factor
ℓ Length of thread engagement, in. (mm)
P Load, lb (kN)
p Pressure applied to thread flank, psi (MPa)
q–y Regression constants
σ Stress [$\sigma = P/(\pi d^2/4)$ for notched or threaded bar; $\sigma = P/(\pi D^2/4)$ for smooth bar], psi (MPa)
v Poisson's ratio
ρ Thread root radius, in. (mm)

Threaded fasteners used in the assembly of structural components are often subjected to high loads that can cause cracking and fracture. Failure processes typically involve crack initiation at a thread root and propagation across the fastener by fatigue or stress corrosion cracking. Final separation occurs when a critical crack size, controlled by the material's toughness, is reached. Accurate stress intensity factor (K) solutions are required to predict crack growth rates and fracture conditions under prototypic loading conditions.

The goal of this study is to calculate nondimensional K values for continuous circumferential cracks emanating from the thread root region of standard thread forms. This crack configuration is applicable to fasteners subjected to uniform membrane stresses, particularly when the material is susceptible to cracking such that multiple cracks initiate around the circumference. These multiple cracks ultimately link together to form a continuous circumferential crack.

In this study, finite-element method (FEM) analyses were performed to develop nondimensional K solutions for standard coarse thread forms. Assumed loading conditions include both remote tension and nut loading (see Figs. 1 and 2), whereby the applied load was reacted at the thread flanks. The effects of thread root radius on K were also evaluated. In addition, K solutions were developed for continuous circumferential cracks emanating from single notches and smooth surfaces. The resulting K solutions were compared with available solutions for threaded fasteners [1–7], notched round bars [8], and smooth bars [8–10].

Numerical Analysis Methods

An FEM program [11] was used to calculate K solutions for continuous cracks in threaded fasteners and notched bars. The program uses eight-node isoparametric quadrilateral elements and a strain energy release rate (G) method to calculate stress intensity factors. The method allows the stress intensity factors to be calculated as a function of crack depth for a given geometry with one computer run. Multiple runs with special crack tip elements being located at different crack depths are not necessary to obtain the stress intensity factor versus crack depth relationship with this strain energy release method.

To calculate the stress intensity factors, the corner and mid-side nodes of a finite element along the specified crack path are released to simulate an increment of crack extension (Δa), equal to the length of the side of the finite element. As nodes are released, nodal displacements and nodal forces along the path are computed. The strain energy released for each crack extension increment is then calculated using the nodal forces prior to the node release and the nodal displacements obtained as a result of the release. Repeated application of this

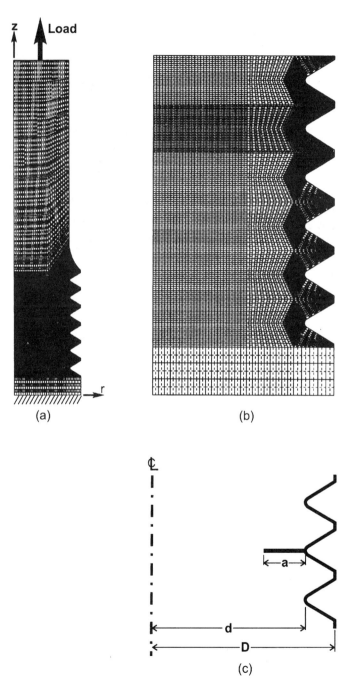

FIG. 1—*Finite-element model of stud subjected to remote tension.* (a) *Axisymmetric model of stud.* (b) *Enhanced detail in threaded region.* (c) *Crack depth measurement.*

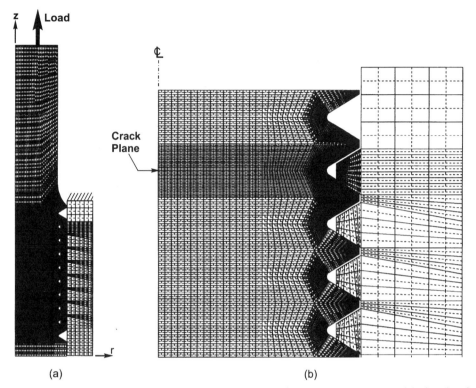

FIG. 2—*Finite-element model of stud subjected to nut loading.* (a) *Axisymmetric model of stud and nut.* (b) *Enhanced detail of thread/nut engagement (all threads are not shown).*

algorithm yields G as a function of the crack depth "a." The corresponding values of K are computed by the following equation

$$K = \sqrt{\frac{EG}{1 - v^2}} \tag{1}$$

where E is the elastic modulus and v is Poisson's ratio.

Four sets of axisymmetric FEM models are developed. The first set represents a uniform cylinder with a continuous circumferential crack. The results from this analysis are compared to literature K solutions [8–10] to verify the analysis method. The second set of models represents notched cylinders with axisymmetric cracks. The notch depths and radii are chosen to represent the standard UNC thread forms listed in Table 1. The notch models provide stress intensity factors for the thread geometry without the complications of multiple threads and thread flank loading.

The third set of models, shown in Fig. 1, represents threaded studs subjected to remote tension. The stud has a reduced shank diameter and six threads are modeled. A uniform axial load is applied to the shank and the stud is fixed vertically at the bottom. The crack plane is located at the root of the second thread from the shank. The fourth set of models, see Fig. 2, represents engaged fasteners. As shown in Table 1, the engaged length is approximately one thread diameter with five to nine engaged threads being modeled. An axial

TABLE 1—*Thread dimensions used in finite-element models. Dimensions are in inches*
(1 in. = 25.4 mm).

| Thread Size | External Thread | | | Internal Thread | | Number of Engaged Threads Modeled |
	Major Diameter	Minor Diameter	Root Radius	Major Diameter	Nut Outside Diameter	
1/4-20UNC	0.24485	0.17725	0.003	0.2500	0.4375	5
1/2-13UNC	0.49305	0.39138	0.006	0.5000	0.7500	7
3/4-10UNC	0.74175	0.61165	0.009	0.7500	1.1250	8
1-8UNC	0.99050	0.82915	0.012	1.0000	1.5000	8
2-4.5UNC	1.98610	1.69560	0.020	2.0000	3.0000	9
4-4UNC	3.98470	3.65604	0.022		Not Modeled	

NOTE—The thread major (and pitch) diameters used in the FEM models are the average of the maximum and minimum values listed in ASME B.1.1-1989. The root radii are based on measurements of a representative sample of fasteners. The minor diameters were calculated to provide a full root radii and to be tangent to the thread flanks.

load is applied to the shank and the top surface of the nut is fixed vertically. The crack plane is located at the first engaged thread. One unengaged thread is located before and after the engaged threads. Figure 3 shows the mesh detail at the thread root. Approximately 7000 quadrilateral elements are in the model. A mesh refinement study showed no significant difference in the K values when using 10 to 30 elements at the thread root. The models used for this study had 20 elements at the cracked thread root. The overlapping stud and nut thread flank nodes were tied together to transmit the thread loads to the nut. This prevents any slippage along the thread flank surfaces.

The axisymmetric models used in this study assume that threaded fasteners consist of a series of parallel notches, rather than a continuous helix. The effect of ignoring the helix shape on the Mode I stress intensity factor is judged to be small, particularly since the helix angles are small.

The results from all four FEM models have been normalized based on the crack depth as measured from the surface of the smooth bar or from the notch or thread root. The notch or thread depth has not been included in the normalization of the results.

Results

Cracks in Smooth Round Bars

The accuracy of the FEM analyses was verified by comparing calculated K solutions for smooth and notched round bars with literature solutions [8–10]. The nondimensional K solution developed in this study for a continuous circumferential crack in a round bar loaded by a uniform far field stress is given by

$$\frac{K}{\sigma\sqrt{\pi a}} = -3.519 + \frac{1.361}{\left(1 - \dfrac{2a}{D}\right)} + \frac{0.0533}{\left(1 - \dfrac{2a}{D}\right)^2} + 10.23\left(1 - \frac{2a}{D}\right)$$

$$- 15.828\left(1 - \frac{2a}{D}\right)^2 + 12.81\left(1 - \frac{2a}{D}\right)^3 - 3.995\left(1 - \frac{2a}{D}\right)^4 \quad (2)$$

where D is the bar diameter and σ the applied stress. The resulting K values are in excellent

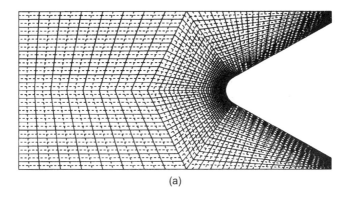

(a)

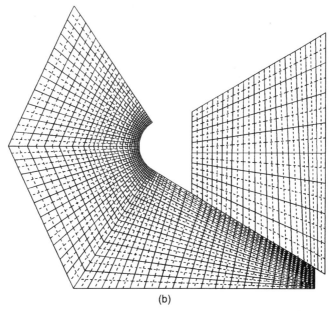

(b)

FIG. 3—*Mesh detail at thread root.* (a) *Mesh detail for one stud thread.* (b) *Enhanced view of thread root showing stud/nut engagement.*

agreement with those developed by Lefort [8] and Tada et al. [9] at all a/D values and Harris [10] at a/D values greater than 0.3. When a/D is less than 0.3, the Harris solution yields K values that are low by about 4%. Figure 4 compares the various K solutions for a continuous circumferential crack in a 1-in. (25.4 mm) round bar subjected to a remote tensile stress of 10 000 psi (68.9 MPa). This geometry was selected for later comparisons with threaded fastener solutions.

Cracks in Notched Round Bars

Nondimensional K solutions for a continuous crack emanating from a single notch in a round bar are provided in Fig. 5. The notch geometries are consistent with the size and

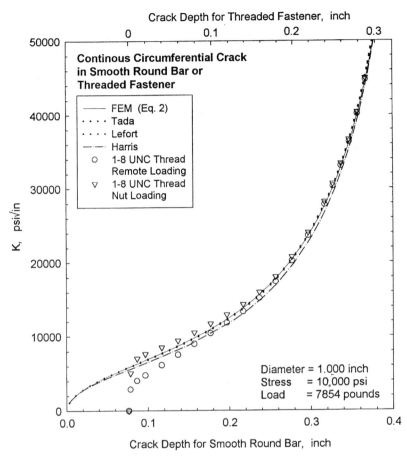

FIG. 4—*K as a function of crack depth for a continuous circumferential crack in a 1-in. (25.4-mm) diameter bar loaded to 7854 lb (34.9 kN), which corresponds to a remote stress of 10 000 psi (68.9 MPa). Values of K calculated by Eq 2 (solid line) are compared with literature solutions (broken and dotted lines)* [8–10]. *Circles and triangles represent K values as a function of crack depth (top axis) for remote-loaded and nut-loaded 1-8 UNC fasteners loaded to 7854 lb (34.9 kN). K values for fasteners can be estimated from smooth bar solutions based on an assumed crack depth (a'—bottom axis) that is equal to the thread depth plus actual crack depth.*

shape of standard UNC thread geometries. The normalized K values are seen to be very high for shallow cracks due to stress concentration effects, drop off rapidly with increasing crack depth-to-minor diameter ratio (a/d), reach a minimum value at an a/d of about 0.1 and increase rapidly beyond an a/d of 0.2 because of the increase in net section stress as the uncracked ligament becomes very small. For a/d values less than 0.2, $K/(\sigma\sqrt{\pi a})$ values are dependent on notch geometry. Notches with the smallest and largest thread dimensions (¼-20 UNC and 4-4 UNC) yield the highest and lowest values, respectively, whereas the other notches (½-13 UNC to 2-4.5 UNC) yield intermediate results. The nondimensional K solutions for each of these data sets are given by the following

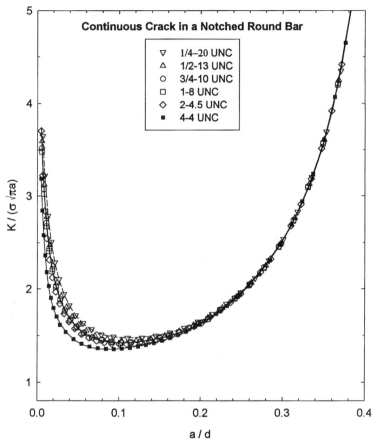

FIG. 5—*Nondimensional* K *solutions for a continuous circumferential crack in a notched round bar. Notch geometries are consistent with UNC thread forms. Correlation equations for* $K/(\sigma\sqrt{\pi a})$ *as a function of* a/d, *as represented by the three lines, are provided in Table 2.*

$$\frac{K}{\sigma\sqrt{\pi a}} = q + re^{-s(a/d)} + t\frac{a}{d} + u\left(\frac{a}{d}\right)^2 + v\left(\frac{a}{d}\right)^3 + w\left(\frac{a}{d}\right)^4 + x\left(\frac{a}{d}\right)^5 + y\left(\frac{a}{d}\right)^6 \quad (3)$$

where

d = minor diameter of notch or thread,
σ = stress acting on notch plane [$\sigma = P/(\pi d^2/4)$],
P = applied load, and
q–y = regression constants.

The regression constants for the notched geometries are given in Table 2.

Figure 6 compares $K/(\sigma\sqrt{\pi a})$ values obtained using Eq 3 and Lefort's solutions for a continuous circumferential crack emanating from a single notch with a 1-8 UNC thread geometry and a 0.012-in. (0.305-mm) root radius. The theoretical stress concentration factor (K_t) for this geometry, which is needed to calculate K per the Lefort solution, was estimated

TABLE 2—*Regression constants for nondimensional* K *solutions for continuous circumferential cracks in notched bars and UNC threaded fasteners.*

Thread Form	q	r	s	t	u	v	w	x	y
NOTCH/Remote Loading (Validity Range: $0.005 \leq a/d \leq 0.4$)									
¼-20	2.9724	2.3701	146.14	−49.168	663.24	−4756.3	19 040.4	−39 186.6	32 963.9
½-13 to 2-4.5	2.6878	2.3931	156.61	−43.165	598.38	−4372.3	17 803.4	−37 149.5	31 623.6
4-4	2.2356	2.6467	199.32	−32.956	483.94	−3651.2	15 267.7	−32 539.5	28 264.2
THREADED FASTENER/Remote Loading (Validity Range: $0.003 \leq a/d \leq 0.4$)									
¼-20	2.1209	1.6351	181.09	−35.837	574.20	−4517.7	19137.2	−40763.3	34960.7
½-13 to 2-4.5	1.7303	1.4640	198.17	−24.232	435.79	−3682.2	16443.9	−36347.5	32073.7
4-4	1.4137	1.5347	299.43	−12.082	253.55	−2366.9	11529.0	−27203.4	25391.2
THREADED FASTENER/Nut Loading (Validity Range: $0.003 \leq a/d \leq 0.4$)									
¼-20 to 2-4.5	3.0149	2.4902	166.26	−51.624	722.92	−5342.9	21757.0	−45123.3	37900.2

$$\frac{K}{\sigma\sqrt{\pi a}} = q + r\, e^{-s(a/d)} + t\frac{a}{d} + u\left(\frac{a}{d}\right)^2 + v\left(\frac{a}{d}\right)^3 + w\left(\frac{a}{d}\right)^4 + x\left(\frac{a}{d}\right)^5 + y\left(\frac{a}{d}\right)^6$$

where a = crack depth, measured from root of notch or thread, and d = minor diameter of notch or thread.

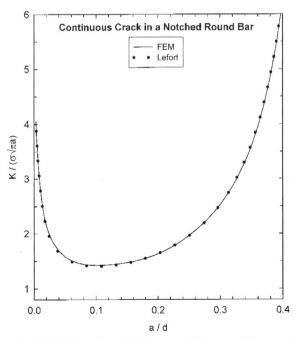

FIG. 6—*Nondimensional* K *solutions for a continuous crack in a round bar containing a single notch with a 1-8 UNC design. FEM solution, per Eq 3, agrees very well with Lefort solution [8].*

to be 4.6 [*12*]. It is seen that the nondimensional K solution given by Eq 3 is in excellent agreement with that developed by Lefort.

Cracks in Threaded Fasteners

Nondimensional K solutions for threaded fasteners with continuous cracks subjected to remote loading are provided in Fig. 7. The general trends are seen to be similar to those described earlier for notched bars. Intermediate size fasteners, with thread sizes ranging from ½-13 UNC to 2-4.5 UNC, exhibit similar K solutions because their thread root geometries scale proportionally. As a result, a single K solution provides an adequate fit of the $K/(\sigma\sqrt{\pi a})$ values for these fasteners. When a/d is less than 0.06, $K/(\sigma\sqrt{\pi a})$ values for the ¼-20 UNC thread are slightly higher, so a separate K solution was developed for this thread geometry. Normalized K values for the 4-4 UNC fastener under remote loading tend to be slightly lower than those for the intermediate sized fasteners at a/d values less than 0.04, so a separate solution was developed for this geometry.

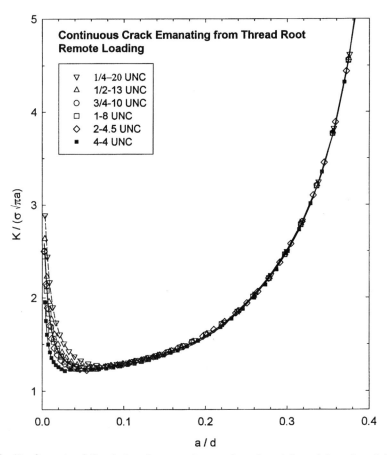

FIG. 7—*Nondimensional* K *solutions for a continuous circumferential crack in a threaded fastener subjected to remote loading. Correlation equations for* K/σ(√πa) *as a function of* a/d, *as represented by the three lines, are provided in Table 2.*

Nondimensional K solutions for threaded fasteners subjected to nut loading, which are provided in Fig. 8, show the same general trends as the remote-loaded notches and threads. The normalized K levels are seen to be independent of thread form at all a/d values. Therefore, a single K solution was used to fit $K/(\sigma\sqrt{\pi a})$ values for nut-loaded threads. Due to model size limitations in the FEM program, an FEM solution for the 4-4 UNC fastener with nut loading could not be obtained.

The nondimensional K solutions for fasteners subjected to remote or nut loading have the same form as Eq 3, and the regression constants are given in Table 2. These solutions are valid for crack depths ranging from 0.3% to 40% of the minor diameter.

The comparison in Fig. 9 of K solutions for remote-loaded and nut-loaded fasteners and remote-loaded notched bars shows some interesting trends. For a/d values greater than 0.25, K solutions are independent of thread and notch geometries and loading conditions, but differences in K solutions are apparent at shorter crack depths. In this regime, nondimensional K values for the remote-loaded threads are significantly lower than those for a single notch, and the minimum value occurs closer to the thread root. This is because the smaller diameter

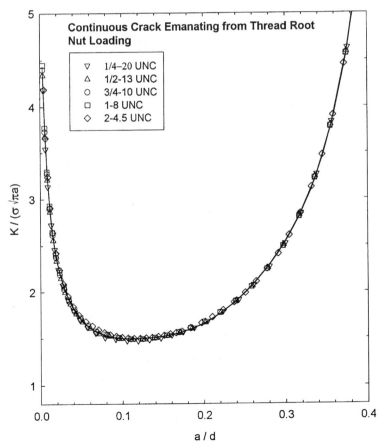

FIG. 8—*Nondimensional* K *solutions for a continuous circumferential crack in a threaded fastener subjected to nut loading. The correlation equation for* K/σ(√πa) *as a function of* a/d, *as represented by the line, is provided in Table 2.*

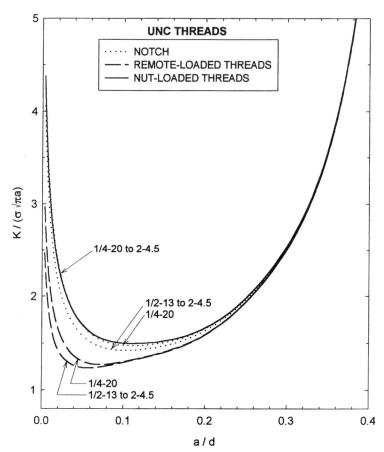

FIG. 9—*Comparison of nondimensional* K *solutions for a notched round bar under remote tension and threaded fasteners under remote tension and nut loading. Regression constants for each curve are provided in Table 2.*

fastener shank and preceding thread shield the crack from the full stress concentration effect of a notch.

The influence of nut loading becomes significant when a/d is less than 0.2, and the magnitude of this effect is seen to increase with decreasing crack depth. At a/d values below 0.02, nut loading increases K by about 50%. It is interesting to note that K solutions for nut-loaded threads and remote-loaded notched bars are similar. By coincidence, it appears that the increase in local stresses due to nut loading is comparable to the stress concentration effect of a single notch.

While Fig. 9 shows that the nut loading effect is significant, it is much lower than that reported by Toribio et al. [4,5], who studied the influence of nut loading on threaded fasteners with surface cracks. They found that nut loading increases K by about 80% to 230% at a/d values between 0.1 and 0.2. While crack geometry differences may partially account for the different observed trends, the biggest effect is believed to be the assumed loading conditions on the thread flanks. Toribio et al. [4,5] assumed a constant pressure (p) applied to the thread flank just below the crack and half of this value ($p/2$) on the second thread flank.

This results in the first thread flank carrying 67% of the total load. They noted that this loading condition was considered to be "conservative from the fracture mechanics viewpoint." Indeed, FEM results from the current study suggest that the assumed loading condition is overly conservative. It was determined from the present analysis of a 1-8 UNC thread that the first two engaged thread flanks carry 28 and 19% of the total load, respectively. These loads are significantly less than those modeled by Toribio et al. [4,5]. Further indication of the conservatism introduced by Toribio et al. [4,5] is provided by consideration of the K values for various ℓ/D ratios. It was found that decreasing the length of engagement (ℓ) causes a modest increase in K when ℓ is less than the major diameter (i.e., $\ell/D < 1$). For example, decreasing the number of engaged threads from eight to four for a $\frac{3}{4}$-10 UNC thread (i.e., decreasing ℓ/D from 1.07 to 0.53) causes a 15% increase in K for an a/D of 0.005 and an 8% increase for an a/D of 0.1. With increasing crack depth, this effect diminishes and it becomes negligible for a/D values beyond 0.3. The sensitivity study also showed that ℓ/D has little effect on K when the length of engagement is greater than the fastener diameter. For example, an 80% increase in ℓ/D from 1 to 1.8 causes less than a 1% decrease in K at an a/D of 0.005 and has a negligible effect for a/D values greater than 0.1. The trends noted above, namely the increase in K with decreasing length of engagement when ℓ/D is less than unity and the lack of a significant effect when ℓ/D exceeds unity, are consistent with good fastener practices where the engaged length should be approximately equal to the nominal fastener size. Note that the nondimensional K solutions given in Table 2 for nut-loaded fasteners were generated for $\ell/D \approx 1$.

Based on these trends, it is concluded that differences in the assumed loading conditions are responsible for the different nut-loading effects reported herein and by Toribio et al. [4,5]. The loading of just the first two threads adjacent to a crack ($\ell/D \approx 0.25$) represents a very conservative stress state that results in overly conservative estimates of K, as noted by Toribio [4]. The present analysis is judged to be slightly conservative because local plasticity effects are expected to produce a more even stress distribution for the first few engaged threads, in comparison with the current linear-elastic analysis where the first engaged thread carries a higher percentage of the load.

Comparison of K *Values for Threaded Fasteners and Smooth Bars*

Figure 4 shows that K solutions for continuous circumferential cracks in smooth cylindrical bars can be used to estimate K for threaded fasteners. This approximation method assumes the following

Bar diameter = major diameter of thread

Assumed crack depth = actual crack depth + thread depth $[a' = a + (D - d)/2]$

Stress = load in shank ÷ area based on major diameter

$$[\text{i.e., } \sigma = P/(\pi D^2/4)]$$

These values are then substituted into any of the K solutions for an axisymmetric crack in a smooth cylindrical bar. Figure 4 compares the approximation method (solid and dotted lines) with the FEM calculated K values for a 1-8 UNC thread (open symbols) subjected to a load of 7854 pounds (34.9 kN). Considering the nut loading case (triangles), the actual K values for cracks with infinitesimal depth approach zero because the $\sqrt{\pi a}$ term is dominant. As the actual crack depth is increased to 0.002 and 0.010 in. (0.05 and 0.25 mm), K increases rapidly to 5000 and 7000 psi$\sqrt{\text{in}}$. (5.5 and 7.7 MPa$\sqrt{\text{m}}$), respectively. These values are seen to straddle the estimated K values, corresponding to the solid and dotted lines. For crack depths between 0.010 and 0.1 in. (0.25 and 2.5 mm), the simple estimation method under-estimates the actual K values by only 10%. For crack depths greater than 0.1 in. (2.5 mm), the estimation method accurately predicts K. In terms of the actual crack depth-to-minor thread diameter ratio, the estimation procedure provides K values that are conservative for very shallow cracks ($a/d < 0.02$), nonconservative by about 10% for a/d values between 0.02 and 0.15, and very accurate for a/d values greater than 0.15.

The same general trends are observed for the remote-loading case (circles), but predictions are substantially more conservative for shallow cracks and remain conservative until a/d values exceed 0.15. For a/d values greater than 0.15, the estimation method accurately predicts K.

Thread Root Radius Effects

A sensitivity study of thread root radius (ρ) effects on K values was conducted for a 1-8 UNC threaded fastener subjected to nut loading and the findings are plotted in Fig. 10. Root radii ranged from 0.003 to 0.024 in. (0.08 to 0.6 mm), with the nominal root radius assumed to be 0.012 in. (0.3 mm). The ratio of K values for a specific root radius versus the nominal 0.012-in. (0.3 mm) radius is plotted in Fig. 10 as a function of normalized crack depth. Root radius has no effect at a/d values greater than 0.02 and very little effect at a/d values between 0.01 and 0.02. This is consistent with Cipolla's findings [1], whereby root radius effects are nearly attenuated when a/d is greater than 0.015.

Figure 10 shows that root radius effects for shallow cracks can be divided into five cate-gories: very sharp ($\rho \approx 0.003$ in. or 0.08 mm), sharp ($\rho \approx 0.006$ to 0.009 in. or 0.15 to 0.23 mm), nominal ($\rho \approx 0.012$ in. or 0.3 mm), blunt ($\rho \approx 0.015$ to 0.018 in. or 0.38 to 0.46 mm), and very blunt ($\rho \approx 0.024$ in. or 0.6 mm). For thread forms other than 1-8 UNC, the root radii associated with these categories are expected to scale proportionally with thread diameter.

The correction factor (F_ρ) for each root radius category is provided by an equation in terms of a/d (see Fig. 10). Thus, values of K for threaded fasteners with various root radii can be computed by the following equation

$$K = F_\rho F \sigma \sqrt{\pi a} \qquad (4)$$

where F is the nondimensional K solution ($K/\sigma(\sqrt{\pi a})$ calculated from Eq 3. F_ρ is equal to unity when a/d is greater than 0.02 and for threads with a nominal root radius, regardless of a/d. For very shallow cracks emanating from thread roots with a non-nominal radius, F_ρ can be calculated from the following equations

Very sharp root radius: $\quad F_\rho = 1 + 0.589\,e^{-377a/d}$ $\qquad (5)$

Sharp root radius: $\quad F_\rho = 1 + 0.175\,e^{-248a/d}$ $\qquad (6)$

Blunt root radius: $\quad F_\rho = 0.774 + 0.226\,(1 - e^{-312a/d})$ $\qquad (7)$

Fully blunt root radius: $\quad F_\rho = 0.667 + 0.333\,(1 - e^{-219a/d})$ $\qquad (8)$

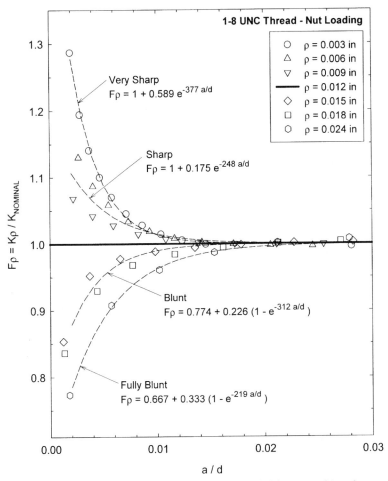

FIG. 10—*Root radius correction factors for a 1-8 UNC threaded fastener subjected to nut loading.*

Comparison of K *Solutions for Threaded Fasteners with Literature Results*

Figure 11 compares the present nondimensional K solutions for a 1-8 UNC threaded fastener containing a continuous crack with literature solutions for fasteners containing continuous, elliptical and straight cracks. For a/d values less than 0.04, all of the K solutions are in good agreement as the literature values are bounded by the nut-loading and remote-loading solutions generated herein. In this regime, the stress concentration effects of the threads overwhelm differences in net section stress associated with continuous versus straight or semi-elliptical surface cracks. The K solution developed by Popov and Ovchinnikov [7] for a continuous crack[3] emanating from the first engaged thread agrees with the present nut-loading solution for a similarly oriented crack.

---

[3] Popov and Ovchinnikov [7] use the term "circular crack," but it appears to be a continuous circumferential crack. They state that the geometry was modeled as a body of rotation, and the K solution used as the basis of their solution is for an axisymmetric crack in a round bar.

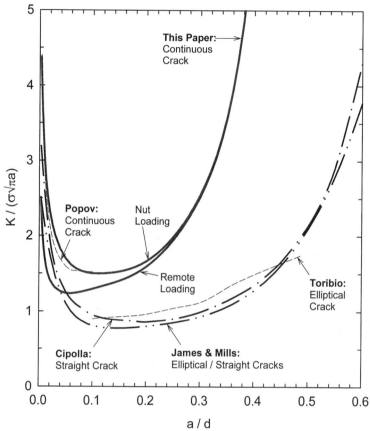

FIG. 11—*Comparison of nondimensional* K *solutions developed by FEM analysis for fasteners subjected to nut loading and remote loading* (*solid curves*) *with literature solutions* (*broken curves*) *[1–3,6,7].*

When a/d is greater than 0.04, K values for continuous cracks are much greater than those for straight or semi-elliptical shaped cracks due to differences in net section stress as the noncracked ligament for a continuous crack quickly decreases with increasing crack depth. This is apparent in Fig. 11 as the K solutions for the two types of cracks diverge with increasing crack depth. It is also seen that the K solutions developed by Cipolla [1] for straight-fronted cracks and by Toribio et al. [3] for semi-circular and semi-elliptical cracks and the composite solution developed by James and Mills [6] are in good agreement.

Conclusions

1. Closed-form nondimensional K solutions were developed for continuous circumferential cracks in UNC threaded fasteners subjected to remote loading and nut loading. The K solutions are valid for a/d values ranging from 0.003 to 0.4 for both loading conditions.

2. An evaluation of thread root acuity effects showed that root radius has no effect on K when the crack depth exceeds 2% of the minor diameter. A root radius correction factor was

integrated into the closed-form solutions that enables K to be computed for a shallow crack emanating from a thread root with an arbitrary radius.

3. For continuous circumferential cracks, it was found that no significant differences exist between remote and nut-loaded studs for a/d values larger than 0.25. For smaller a/d values, the increase in stress at the thread root caused by the thread flank loading increases the K values for the nut-loaded case. At an a/d value of 0.05, the nut-loaded K values are 60% larger.

4. Comparison of the K solutions for a round smooth bar with an effective crack depth that included the thread depth with the FEM solution for a nut-loaded thread showed excellent agreement for a/d values larger than 0.15. The round bar solution is nonconservative by about 10% for a/d values between 0.02 and 0.15.

Acknowledgment

This work was performed under U.S. Department of Energy Contract DE-AC11-98PN38206 with the Bettis Atomic Power Laboratory.

References

[1] Cipolla, R. C., "Stress Intensity Factor Approximations for Cracks Located at The Thread Root Region of Fasteners," *Structural Integrity of Fasteners, ASTM STP 1236,* American Society for Testing and Materials, West Conshohocken, PA, 1995, pp. 108–125.

[2] Cipolla, R. C., "Stress Intensity Factor Approximations for Semi-elliptical Cracks at the Thread Root of Fasteners," *Improved Technology for Critical Bolting Applications,* ASME MPC-Vol. 26, pp. 49–58.

[3] Toribio, J., Sanchez-Galvez, V., Astiz, M. A., and Campos, J. M., "Stress Intensity Factor Solutions for a Cracked Bolt Under Tension, Bending and Residual Stress Loading," *Engineering Fracture Mechanics,* Vol. 39, 1991, pp. 359–371.

[4] Toribio, J., Sanchez-Galvez, V., and Astiz, M. A., "Stress Intensification in Cracked Shank of Tightened Bolt," *Theoretical and Applied Fracture Mechanics,* Vol. 15, 1991, pp. 85–87.

[5] Toribio, J., "Stress Intensity Factor Solutions for a Cracked Bolt Loaded by a Nut," *International Journal of Fracture,* Vol. 53, 1992, pp. 367–385.

[6] James, L. A. and Mills, W. J., "Review and Synthesis of Stress Intensity Factor Solutions Applicable to Cracks in Bolts," *Engineering Fracture Mechanics,* Vol. 30, 1988, pp. 641–654.

[7] Popov, A. A. and Ovchinnikov, A. V., "Stress Intensity Factors for Circular Cracks in Threaded Joints," *Strength of Materials,* Vol. 15, 1983, pp. 1586–1589.

[8] Lefort, P., "Stress Intensity Factors for a Circumferential Crack Emanating from a Notch in a Round Tensile Bar," *Engineering Fracture Mechanics,* Vol. 10, 1978, pp. 897–904.

[9] Tada, H. P., Paris, P. C., and Irwin, G. R., *The Stress Analysis of Cracks Handbook.* Del Research Corporation, St. Louis, MO, 1985.

[10] Harris, D. O., "Stress Intensity Factors for Hollow Circumferentially Notched Round Bars," *Journal of Basic Engineering,* Transactions of the ASME, Series D, Vol. 89, 1967, pp. 49–54.

[11] Friedman, E., "Curvature Effects in Thick Cylindrical Shells with Continuous Surface Cracks," *Proceedings, 5th International Conference on Pressure Vessel Technology—Vol. II: Materials and Manufacturing,* ASME, New York, 1984, pp. 761–776.

[12] Peterson, R. E., *Stress Concentration Factors,* Wiley and Sons, 1974.

Analysis Techniques

D. Barke,[1] W. K. Chiu,[1] and S. Fernando[2]

Residual Strength Assessment of Stress Corrosion in High Strength Steel Components

REFERENCE: Barke, D., Chiu, W. K., and Fernando, S. **"Residual Strength Assessment of Stress Corrosion in High Strength Steel Components,"** *Structural Integrity of Fasteners: Second Volume, ASTM STP 1391,* P. M. Toor, Ed., American Society for Testing and Materials, West Conshohocken, PA, 2000, pp. 105–119.

ABSTRACT: Testing of finished high strength steel components for their susceptibility or resistance to stress corrosion cracking (SCC) provides data of immense value to designers and manufacturers. Reported in the literature are many different techniques of testing for SCC which have varying advantages and disadvantages. However, most techniques have been questioned on the grounds of applicability and reproducibility.

A test which rapidly ranks similar components, in order of resistance to SCC, is in demand. Techniques which quantify the extent of damage, after a period of time, such as the breaking load method, are able to overcome some of the limitations of other SCC tests. The use of the breaking load method has been reported to provide reliable results in the comparison of certain aluminum alloys.

This paper applies the breaking load method, a residual strength test, to high strength steel components. Bolts, rather than tensile test specimens, were used, to enable the effect of geometry and manufacturing methods on susceptibility to SCC to be studied. It is concluded that the breaking load method can be used to rank differences between materials, manufacturing methods and geometry.

KEYWORDS: high strength steel, breaking load method, stress corrosion cracking, fasteners, stress corrosion cracking test methods

The aim of establishing a simple stress corrosion cracking (SCC) test is to accurately predict the long-term resistance to failure of a component, such as a bolt, in its service environment. Regretfully, there is no single parameter which defines the resistance of a material to stress corrosion cracking in an environment, so the results from one test are not directly comparable with another. Each method has benefits and drawbacks, which should be considered when developing a test. One technique that shows some promise is the residual strength test, which quantifies the damage in a test specimen, after a period of exposure.

One type of residual strength technique is the breaking load method ASTM Standard Test Method for Determining Stress-Corrosion Cracking Resistance of Heat Treatable Aluminum Alloy Products Using Breaking Load Method (G 139-96) which exposes specimens to a constant load or total strain in an environment which will induce SCC damage, for a brief period of time (typically two to ten days for aluminum, e.g., ASTM G 139). Subsequent to

---

[1] Graduate student and associate professor, respectively, Department of Mechanical Engineering, Monash University, Clayton 3168 Victoria, Australia.
[2] Project leader, Ajax Technology Centre, Locked Bag 6000, Malvern 3144 Victoria, Australia.

this period the strength remaining in the component is obtained by loading the specimen to failure in tension. Analysis of the residual strength by the Box-Cox statistical transform enables accurate measurement of the material behavior. This technique has been successfully used to rank aluminum alloys and is claimed to be applicable to other materials [1], although no materials other than aluminum are reported in the literature.

The breaking load method would appear to overcome some of the weaknesses of other SCC tests [1], such as those of constant parameter tests (the parameter being either load or strain) stress a specimen by either two-, three- or four-point bending or uniaxial tension. Turnbull has claimed that constant total strain specimens behave in a similar manner to the components they model [2]. In either constant parameter loading mechanism, some specimens will not fail because they are completely inert to the environment, while others may crack, but not break in two in the test period, and still others may fail for reasons other than SCC. Interpretation of the results then becomes difficult and is confused by the finite period of the test [1].

The slow strain rate test overcomes the nonfailure problem of constant parameter tests by increasing the strain on the specimen with time. This inevitably causes failure in the specimen either by ductile fracture due to overload or by SCC. Any specimen exposed in a slow strain rate test must be examined metallographically, to determine the mode of failure. Slow strain rate tests may promote SCC in materials that do not otherwise exhibit SCC [3]. It is necessary to judge which materials are then acceptable for service [4], making such tests subjective rather than objective. Despite being developed as an ad-hoc method of testing, it has been used in mechanistic studies, since strain rate is a key parameter in determining failure [4].

The shortcomings of constant parameter tests appear to be overcome by the breaking load method. However, if a specimen is broken before it can be tensile tested, the method cannot provide a result. Statistical techniques of overcoming this problem have been proposed [5].

This paper shows that the breaking load method, as defined in ASTM G 139, can be used to measure the SCC resistance of high strength steel components such as bolts. By testing a mass-produced component, the effect of manufacturing processes, material and geometric features on SCC resistance of the component can be evaluated. The paper also highlights the rapid evaluation available from this test.

Experimental Background

Stress/strain levels present must be quantified in an SCC test for it to be properly characterized. Failure may be accelerated by introducing a stress concentration. An example of this is loading the test specimen in bending (in a similar manner to that proposed by the ASTM Standard Practice for Preparation and Use of Bent-Beam Stress-Corrosion Test Specimens (G 39). This will facilitate a simple test frame design and also overcomes torsion associated with loading bolts by standard means.

To determine how this loading mechanism would affect the crack growth in the specimen when SCC occurs, two computer simulations were undertaken using the Pafec version 5.1 finite-element analysis package. The nature of the problem dictated that the model be constructed from 3-D elements. The crack is located in the region of highest stress where SCC is most likely. The length of the model was 6.0 times the diameter. Crack lengths used in the model were 0.2, 0.3 and 0.4 times the model diameter, and the crack was located a distance 0.5 times the diameter away from one end, which was fixed at all degrees of freedom.

Analysis 1

To show the effect of bending in constant strain upon crack growth, a model was given a displacement 0.4 times its diameter at the right-hand (free) end, as seen in Fig. 1a, which is similar to that given to the test specimen, as seen in Fig. 2. The stress intensity factor K_L, a measure of the stress singularity at the crack tip in the structure [6], was evaluated using the formula:

$$K_L = \frac{Ev}{4} \sqrt{\frac{2\pi}{L\left(1 - \frac{L}{2a}\right)}} \tag{1}$$

where

E = Young's modulus of the alloy.
v = vertical displacement measured at a point along the crack face is a distance L from the crack tip, and
a = crack length [7].

The normalized stress intensity factor was calculated at the various crack lengths (Table 1). The normalized stress intensity factor is defined as:

$$K_{\text{normalized}} = K_L / K_2 \tag{2}$$

where $K_{\text{normalized}}$ is the normalized stress intensity factor, K_L is the stress intensity factor when crack length is L and K_2 is the stress intensity factor when crack length is 0.2 times the diameter of the model.

It is shown (Table 1) that the stress intensity factor during crack growth varies by 10 to 20%, which is within the range of experimental error. It can be concluded that stress intensity is relatively insensitive to crack length and therefore load levels are unlikely to vary significantly during the test.

Analysis 2

Estimation of residual strength by loading the specimen to failure in uniaxial tension, as per the breaking load method which requires the specimen to be removed from the environment and pulled to failure in a tensile test machine. For this technique to be useful, it is important that an appreciable difference in residual load of the various cracked specimens is obtained.

Here a model with various crack lengths was subjected to a given tensile load along the axis of the bolt, as shown in Fig. 1b. The stress intensity factor has been evaluated using Eq 1 and the normalized stress intensity factor evaluated using Eq 2.

The difference in normalized stress intensity factor (Table 2) shows that the residual load of the bolt can provide an indication of the extent of SCC. As an example, a change of 54% in the load capacity of the component is expected when the normalized crack length grows from 0.2 to 0.3.

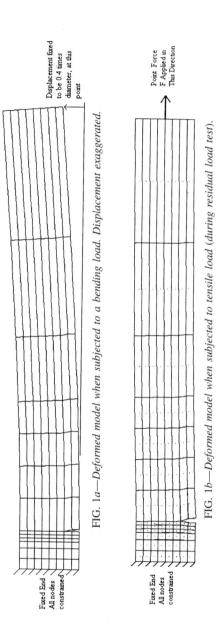

FIG. 1a—Deformed model when subjected to a bending load. Displacement exaggerated.

FIG. 1b—Deformed model when subjected to tensile load (during residual load test).

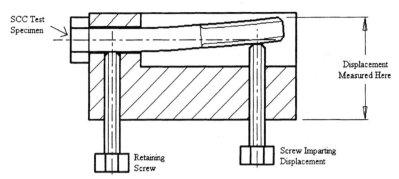

FIG. 2—*A cross-sectional view of the test frame, showing the manner in which bolts were secured in the frame and the means of bending the bolt. This drawing is to scale. Displacement is defined as the change in the distance measured between the top of the bolt and the bottom of the frame.*

Summary of Simulation Results

From Analysis 1, it can be concluded that changes in crack length in a component loaded in constant displacement bending are not likely to significantly affect crack growth characteristics. Analysis 2 shows that a specimen exposed to a known load will undergo significant increases in stress intensity with increasing crack growth. It can be concluded that the failure load can be used as an indicator of SCC damage for this geometry.

Experimental Procedure

Overview of the Breaking Load Method and Necessary Conditions for Such a Test

The breaking load method is a two-parameter SCC test. Each specimen of a group will be exposed to one of n strain levels for one of m time periods, making a total of $n \times m$ groups of specimens. The levels and periods used allow a significant stress corrosion crack to grow without it causing complete failure of the specimen. For the entire period that the specimen is exposed to an environment which causes SCC, each specimen is loaded at a constant strain. Subsequent to this period of exposure under strain, every specimen is removed from its aggressive environment and loaded to failure in a tensile test. The maximum load at which failure occurs divided by the original area of the test specimen is defined as the residual strength of the component.

In order to undertake the breaking load method, it is necessary to know the tensile properties of the component including the ultimate strength in order for degradation to be iden-

TABLE 1—*Normalized stress intensity factor due to constant bending displacement.*

Normalized Crack Length (Ratio of Crack Length to Diameter)	Normalized Stress Intensity Factor, K_L (Defined in Eq 2)
0.2	1.00
0.3	1.10
0.4	1.16

TABLE 2—*Normalized stress intensity factor resulting from tensile load.*

Normalized Crack Length (Ratio of Crack Length to Diameter)	Normalized Stress Intensity Factor, K_L (Defined in Eq 2)
0.2	1
0.3	1.54
0.4	2.54

tified. SCC must be present, and the minimum stress/strain and shortest exposure time to induce SCC failure must be known. It is also necessary to characterize the test frame used to load specimens in order to quantify exposure strain levels.

Component Tensile Properties

In order to determine the tensile properties of the bolts used in this test, samples were strained to failure at a constant strain rate of 0.01 mm/s. Load was recorded using a load cell and strain was recorded by fixing a strain gage to the flat surface of the bolt. All tensile tests in this paper were undertaken with an MTS tensile testing machine, digitally controlled by a TestStar Controller.

Damage Due to Bending Level

A number of specimens were also bent to a known displacement in a test frame (shown schematically in Fig. 2). Immediately after achieving this displacement, the specimens were removed from the frame and pulled to failure by loading the bolt in uniaxial tension to determine whether any static damage had occurred due to bending. In every case in this paper, displacement was defined as the vertical dimension shown in Fig. 2 and was measured using a digital Vernier caliper accurate to 0.01 mm.

Effect of Sodium Chloride on Bolts

A constant strain test was used to establish the presence of SCC. A group of nine bolts was mounted in test frames, shown in Fig. 2, and each displaced by amounts of 1, 5 and 8 mm. These bolts were placed in an environment of 3% sodium chloride (NaC1), at room temperature. Simultaneously, a control group was mounted in test frames, displaced by amounts identical to those in the first group and placed in an airtight drum filled with silica gel crystals. Both groups of specimens were left in their respective environments for 14 days and inspected daily. Subsequent to this period, each test specimen was removed from the test frame. This test established the strain levels and time periods which would cause significant SCC damage to the component.

Test Frame Characterization

The test frame, shown schematically in Fig. 2, was characterized by placing a strain gage on the upper surface of the bolt and recording strain at the fulcrum versus displacement. This process was repeated six times using different frames. Each of the displacement-strain curves was approximated with a fifth-order polynomial. A mean of the polynomials was taken and subsequently used to estimate exposure strain, as a function of displacement.

The Breaking Load Method (Details of the Procedure Used in this Experiment)

Having established that SCC was present, three displacement levels (3, 4.5 and 6 mm) and three exposure periods (5, 10 and 20 h) were selected to be used in the breaking load method.

For each of the nine test conditions, nine bolts were strained in test frames (as shown schematically in Fig. 2) by bending to a selected displacement. The entire frame and strained bolt were immersed in 3% sodium chloride solution for a selected period of time. Immediately subsequent to exposure in sodium chloride, the retaining screw and the screw imparting bending upon the bolts were released and the SCC test bolt removed from the frame.

To determine the residual strength after exposure, each specimen was gripped by the head and thread and loaded in uniaxial tension, until failure occurred. The load rate was fixed at 1 kN/s. The maximum load withstood by the exposed bolt was recorded. This load, divided by the area of the bolt (at the thread root diameter) was deemed to be the residual strength of the component. Where failure occurred before the component could be tensile tested, the residual strength was estimated in accordance with ASTM G 139.

Residual strength results were statistically transformed to a normal distribution by the Box-Cox method, as defined in ASTM G 139 and the mean and standard deviation of the transformed values were evaluated. The Box-Cox transform is based on the assumption that the data can be altered by a function of the form:

$$y^{(\lambda)} = k + l_{(\lambda)}d^{\lambda} \tag{2}$$

where d is the datum recorded in the environment; k, λ and 1 are constants determined by the method defined in ASTM G 139; and $y^{(\lambda)}$ is the transformed point.

The transform requires no knowledge of the initial distribution of data points d and, since $y^{(\lambda)}$ is normally distributed, analysis using standard statistical techniques is possible [8].

Results

Component Tensile Properties

The tensile properties of the bolt were established by testing seven specimens to determine the ultimate strength (Table 3) and the stress-strain characteristics, of which a typical curve

TABLE 3—*Ultimate tensile strength of a sample of bolts.* *

Tensile Specimen	Ultimate Strength (kN)	Ultimate Strength (MPa)
1	86.027	1140.4
2	86.665	1148.9
3	87.363	1158.2
4	73.533	974.8
5	75.799	1004.8
6	74.145	982.9
7	72.802	965.17
average	79.47	1053.6

* Bolts tested are identical to those used in SCC testing.

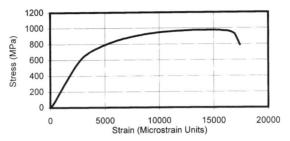

FIG. 3—*A typical stress-strain curve for the high strength steel bolt used in this SCC test.*

is shown in Fig. 3. Failure of the bolt occurred between a strain of 13 000 and 17 500 (microstrain units).

Test Frame Characterization

The output of the strain gage mounted near the fulcrum of the bolt was converted to equivalent strain at the fulcrum and logged against displacement (measured as shown in Fig. 2, using a digital vernier caliper) at the tip of the bolt for six sample bolts. The resulting strain at the fulcrum versus displacement at the bolt tip was plotted (Fig. 4). Fifth-order polynomial curves were fitted to the results obtained. The average curve of those polynomials is also plotted (Fig. 4).

Damage Level Due to Bending

Three specimens were bent in a test frame to a displacement of 6.00 mm, as shown in Fig. 2, and left in a desiccated environment for a period of 1.5 h. Subsequently, the specimens were removed and pulled to failure. Failure loads (Table 4) are within the range of ultimate strengths observed (refer to Table 3).

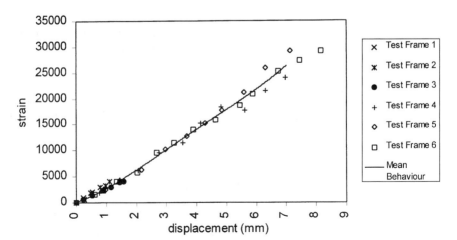

FIG 4—*Strain vs. displacement: fastener deflection.*

TABLE 4—*Tensile strengths of bolts tested immediately after bending to a displacement of 6.00 mm.**

Bent Specimen	Failure Load (kN)	Failure Load (MPa)
1	82.796	1097.65
2	82.262	1090.58
3	84.123	1115.25

* Specimens subjected to this test were not immersed in sodium chloride for any period of time.

Effects of Sodium Chloride on Bolts

Specimens loaded in frames (see Fig. 2) were observed to fail within 72 h (Table 5) when exposed to 3% sodium chloride at room temperature. Visual inspection of identically stressed specimens having been exposed for 14 days in a desiccated environment identified no cracks.

Results of the Breaking Load Method

It can be seen in a plot of behavior of specimens against time and displacement (Fig. 5a) that there is a trend of decreasing component strength with time. Mean trends accentuate this behavior.

Box-Cox Analysis of Behavior

Having obtained data of exposure strain and residual strength, analysis was undertaken using the Box-Cox statistical transform, as defined in ASTM G 139. Residual strength was transformed into the Box-Cox metric. Figure 5b plots this behavior.

Fracture Surfaces

Figure 6a shows a typical failure surface from a tensile test to determine a stress-strain curve. The surface is at 45 deg to the axis of the bolt and follows a weakness in the component geometry. It can also be seen that the surface is consistent with ductile fracture.

TABLE 5—*Results of the time to failure test in an aggressive and control environment.*

Deflection (mm)	Environment	Time to Failure	Deflection (mm)	Environment	Time to Failure
0.99	3% NaCl	no failure*	0.97	desiccated	no failure*
0.99	3% NaCl	no failure*	1.00	desiccated	no failure
1.58	3% NaCl	no failure*	1.59	desiccated	no failure
4.57	3% NaCl	F.U.R.†	4.59	desiccated	no failure
4.76	3% NaCl	3 days	4.77	desiccated	no failure
5.53	3% NaCl	F.U.R.	5.60	desiccated	no failure
8.04	3% NaCl	F.U.R.	8.00	desiccated	no failure
8.19	3% NaCl	F.U.R.	8.10	desiccated	no failure
8.23	3% NaCl	3 days	8.25	desiccated	no failure

* No failure was observed within 14 days, after which the components were removed. No observable surface cracks were present.
† F.U.R. = Failure upon removal; exposure period was 14 days. Fracture surface consistent with an SCC crack 90% of the way through the bolt.

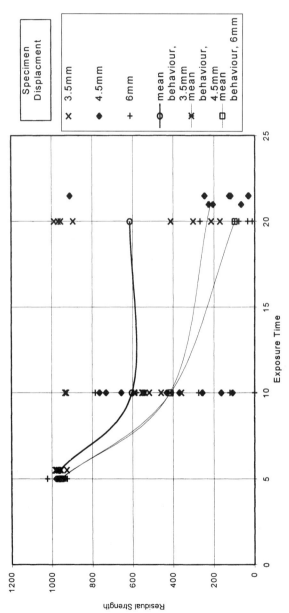

FIG. 5a—Time and displacement-dependent behavior of specimens immersed in 3% NaCl.

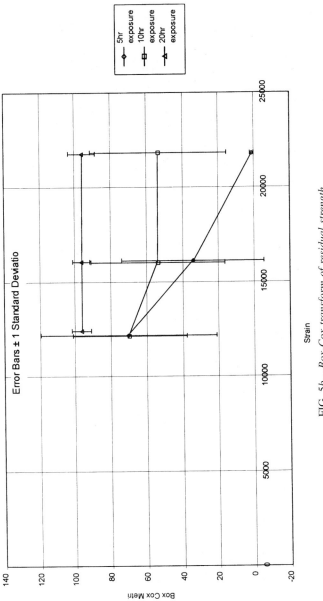

FIG. 5b—*Box-Cox transform of residual strength.*

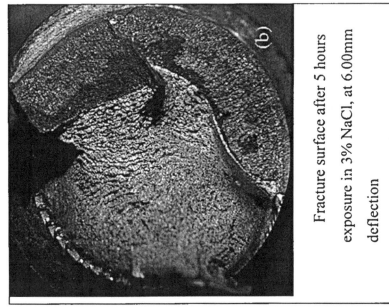

Fracture surface after 5 hours exposure in 3% NaCl, at 6.00mm deflection

The fracture surface of a bolt subjected to a tensile test.

FIG. 6—*Representative bolt fracture surfaces after testing.*

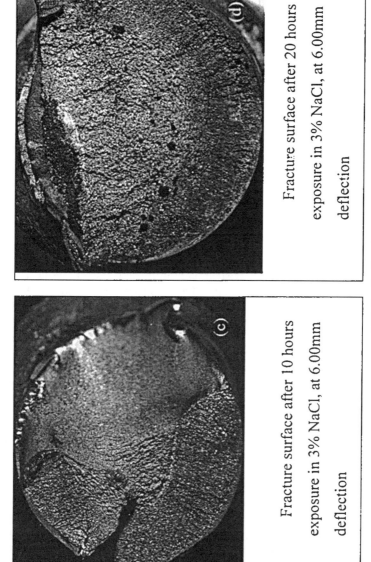

Fracture surface after 20 hours exposure in 3% NaCl, at 6.00mm deflection

Fracture surface after 10 hours exposure in 3% NaCl, at 6.00mm deflection

FIG. 6—Continued.

Similar fracture surfaces were observed when bolts were subjected to a 6.00 mm displacement in bending and subsequently pulled to failure, which confirms the results obtained in the time to failure test, establishing the effects of sodium chloride on the bolt (Table 5). This surface can be compared with those of specimens subjected to the breaking load method SCC test.

Figure 6b shows the fracture surface of a bolt after 5 h exposure. SCC cracks can be seen on the upper right-hand side and on the bottom left of the picture. These cracks are on different planes approximately 2 mm apart. Tensile failure occurred at a load of 72.33 kN, which is slightly less than the mean strength of 79.47 kN. The depth of the upper right SCC crack is approximately 2.57 mm, while the lower right crack is approximately 1.77 mm deep.

Figure 6c shows a similar specimen, exposed for 10 h to sodium chloride. Again, two SCC cracks can be observed in the upper and lower left of the picture, this time with maximum depths of 1.68 and 2.92 mm. This is reflected in a failure load of 59.25 kN.

Figure 6d shows a specimen immersed for 20 h in sodium chloride. The SCC crack has grown 90% of the way through the specimen and ends at the top of the darkened section; the length is approximately 7.90 mm. Failure occurred under a load of 5.954 kN.

Discussion

The results presented show that SCC is present when the specimens are exposed to high levels of strain in an aggressive environment of 3% sodium chloride solution at room temperature. Despite the high levels of strain used in the test, loss of tensile strength was not observed when specimens were not exposed to the aggressive environment. In loading these unexposed specimens to failure, fracture surfaces were observed to be consistent with those that had not been strained in bending or exposed to the aggressive environment, prior to tensile testing.

Further to this, it has been possible to observe that the Box-Cox analysis provides a significant means of measuring the behavior of SCC-affected components. The results appear similar to those presented in ASTM G 139. In Fig. 5b, it is seen that increasing exposure strain and exposure time will decrease the Box-Cox metric.

When the data have been Box-Cox transformed to a normal distribution, it is possible to compute an average and a variance of the result. This is plotted in Fig. 5b. The majority of the data will lie between plus or minus one standard deviation of the mean. Despite the large error level shown at some points when the results of the test at the maximum strain level are considered, it is evident that there is a significant change in Box-Cox metric between a 5 h exposure and a 20 h exposure. The error bars would overlap only if they were extended by greater than plus or minus three standard deviations, or 98% of the data [9].

The use of bolts, rather than tensile specimens, in this test program indicates that it is possible to test SCC susceptibility, using the breaking load method, with generic component structures. By loading the bolts in bending rather than in tension, the exposure strain can be quantified. To this end, the technique is simple and more accurate than using a torque wrench to tension the bolt, as well as being cheaper and quicker than strain gaging every specimen.

Computer simulation of the loading mechanism geometry shows that stress intensity at the crack tip does not vary significantly. The simulation also shows that the tensile load required to pull a bolt to failure will decrease as the crack grows from 20 to 40% of the way through the specimen.

Conclusion

It has been clearly shown that there is a relationship between material, exposure strain and length of exposure time, where SCC is present. This relationship can be tested by determining the load bearing capacity of the component. This residual strength can be used as a means of quantifying exposure damage that occurs in a component. By testing a component, rather than a tensile test specimen, the effects of materials, machining processes and geometry on SCC resistance on the component can be observed.

While the loading mechanism used in this test does not exactly replicate that which the component would undergo in service, the mechanism has been chosen because it allows load levels to be applied easily, and efficiently quantifies the strain level of every specimen. Computer simulation has shown the change in stress intensity under this loading mechanism, as crack growth occurs, to be negligible.

It is shown that component load capacity, or residual strength, is a function of time-in-environment and load level, and that this can be observed when the time/load level is plotted against the residual strength. By statistically transforming the residual strength, using the Box-Cox method, component behavior is more clearly observed. Since the transformed data are normally distributed, standard statistical techniques can be subsequently applied, allowing confidence levels to be estimated.

References

[1] Sprowls, D. O., "Evaluation of Stress Corrosion Cracking," *Stress Corrosion Cracking; Materials Performance and Evaluation*, R. H. Jones, Ed., ASM International, Materials Park, OH, 1992, pp. 363–415.
[2] Turnbull, A., "Test Methods for Environment Assisted Cracking," *British Corrosion Journal*, Vol. 27, No. 4, 1992, pp. 271–289.
[3] Parkins, R. N., "Development of Strain-Rate Testing and its Implications," *The Slow Strain Rate Technique, ASTM STP 665*, G.M.a.P., J. H. Ugiansky, Ed., 1979, American Society for Testing and Materials, West Conchohocken, PA, pp. 5–25.
[4] Payer, J. H., Berry, W. E., and Boyd, W. K., "Constant Strain Rate Technique for Assessing Stress-Corrosion Susceptibility," *Stress Corrosion—New Approaches*, H. L. Craig Jr., Ed., American Society for Testing and Materials: West Conshohocken, PA, 1976, p. 419.
[5] Colvin, E. L. and Emptage, M. R., "The Breaking Load Method: Results and Statistical Modification from the ASTM Interlaboratory Test Program," *New Methods for Corrosion Testing of Aluminum Alloys, ASTM STP 1134*, V. S. Ugiansky and G. M. Agarwala, Eds., American Society for Testing and Materials: West Conshohocken, PA, 1992, pp. 82–100.
[6] Broek, D. W., *Elementary Engineering Fracture Mechanics*, 4th ed., Kluwer Academic Publishing Group, AH Dordrecht, 1986, p. 501.
[7] Jones, R. and Williams, J., "An Introduction to Fracture Mechanics," *Transactions of the Institute of Engineers, Australia; Mechanical Engineering*, Vol. 14, No. 4, 1989, pp. 185–223.
[8] Box, G. E. P. and Cox, D. R., *An Analysis of Transformations*, Vol. 2, 1964, pp. 211–252.
[9] Walpole, R. E. and Myers, R. H., *Probability and Statistics for Engineers and Scientists*, 5th ed., Maxwell Macmillan International, Sydney, 1993, p. 766.

Ronald J. Hukari[1]

Thread Lap Behavior Determination Using Finite-Element Analysis and Fracture Mechanics Techniques

REFERENCE: Hukari, R. J., **"Thread Lap Behavior Determination Using Finite-Element Analysis and Fracture Mechanics Techniques,"** *Structural Integrity of Fasteners: Second Volume, ASTM STP 1391,* P. M. Toor, Ed., American Society for Testing and Materials, West Conshohocken, PA, 2000, pp. 120–132.

ABSTRACT: A new aerospace bolt thread lap inspection criterion is proposed based on fatigue crack propagation trajectories predicted by finite-element analysis (FEA) and fracture mechanics techniques.

Fastener producers and end users incur high costs due to thread lap problems. Current thread lap inspection criteria are 30 to 40-years-old, and are based on an intuitive understanding of potential thread lap behavior. These inspection criteria are ambiguous and make inspection difficult. Some costs can be avoided and quality can be improved by redefining acceptable and unacceptable thread laps. Specifically, these inspection criteria can be improved by using fracture mechanics and FEA techniques developed in the last 30 years.

Two-dimensional, axisymmetric, full nut-bolt-joint geometry, FEA models with elastic-plastic material properties, and contact elements at the thread interfaces were analyzed using ANSYS FEA software. Singular crack tip elements were used. Laps of expected geometries were built into the models, and then assumed to propagate as fatigue cracks. Cracks were iteratively grown using the maximum tangential principal stress criterion to determine the direction of crack propagation. Certain types of thread laps were shown to propagate in such a way that bolt failure would not occur. Conversely, other types of laps were predicted to propagate in such a way that bolt failure would be expected.

A simpler, less subjective criterion for inspecting laps is presented. This criterion takes advantage of the benign laps. The proposed method allows any lap contained entirely within a particular zone to be acceptable.

KEYWORDS: fracture mechanics, finite-element analysis, bolt thread, thread lap, fatigue crack trajectory, fatigue crack path, maximum tangential principal stress criterion, inspection, mixed mode fracture, aerospace fastener, automotive fastener

Nomenclature

β Crack initial inclination angle relative to loading direction.

ε Strain (plastic).

K_I Stress intensity factor for Mode I loading.

K_{II} Stress intensity factor for Mode II loading.

ΔK_I Alternating stress intensity factor for Mode I loading.

ΔK_{II} Alternating stress intensity factor for Mode II loading.

θ_C Angle between initial crack and direction of crack propagation.

$\tau_{r\theta}$ Shear stress relative to polar coordinate system.

---

[1] Product engineer, SPS Technologies, Jenkintown, PA 19046.

Background

Thread Laps

Bolts requiring high fatigue strength typically have threads formed by rolling. Flaws called laps can be formed during the rolling operation. A lap is typically defined in fastener industry documents as a "surface defect, appearing as a seam, caused by folding over hot metal or sharp corners and then rolling or forging them into the surface but not welding them."

Thread lap problems cost fastener manufacturers millions of dollars a year. This cost is ultimately passed on to end users in the form of higher price, delayed delivery, or loss of service. Some of this cost can be avoided by improving and updating thread lap inspection criteria.

Improvement can be made both with content and presentation of the criteria. Current thread lap inspection criteria are 30 to 40-years-old, and based on an intuitive understanding of how thread laps might behave. Their success in preventing thread failures is likely due to their overall conservatism rather than their accuracy in reflecting the mechanics of potential thread lap behavior. Inconstancy exists between different specifications. In some specifications permissible laps may potentially be more dangerous than certain nonpermissible laps. The inspection criteria can be improved by using fracture mechanics and finite-element analysis (FEA) techniques developed in the past 30 years.

Thread lap inspection criteria have a common format in aerospace and automotive industry documents. The criteria typically consist of a sketch as shown in Fig. 1. This figure is condensed from SPS Technologies thread inspection document SPS-I-650. Here, two thread profiles are shown with the location of the nonpressure (nonbearing) thread flank and basic thread dimensions indicated. Superimposed on one profile is a set of sketched lines that indicate permissible lap geometries and on the other profile is a set of nonpermissible sketched lines. In some documents a table accompanying the sketch gives allowable lap lengths for different thread geometries and materials.

Even the best criteria are subjective, ambiguous, and difficult to inspect under manufacturing conditions. Only a small portion of all possible lap geometries are indicated. No explanation is provided for interpreting the gray area between permissible and nonpermissible geometries. The sketched lines do not photocopy well; a second- or third-generation photocopy, which an inspector may be referencing, may not reflect well what was on the original sketch.

Although similar in appearance to a crack, a lap is different from a crack. Most importantly, at the tip of a lap (the end embedded in the material) the material is folded around the tip and is continuous. Thus it would not be expected that the singularity that exists at a crack tip would exist at the tip of a lap. For typical bolting materials it is very possible that compressive residual stresses are induced at the tip of the lap by the folding of material.

Fracture Mechanics Background

The maximum tangential stress (MTS) criterion for predicting mixed mode crack trajectory (θ_C) was proposed in 1963 [1]. Since then a number of criteria have been proposed and studied [2–4], among them the maximum tangential principal stress (MTPS) criterion [5,6]. The MTPS criterion is based on the theory that the crack will propagate in a direction perpendicular to the maximum principal stress at the crack tip. Thus the crack extends along a radial direction given by $\tau_{r\theta} = 0$, for a polar coordinate system with its origin at the crack tip. The MTPS criterion is sometimes referred to as the zero shear stress criterion. Application of this criterion is easily accomplished analytically and with FEA. It is similar but not identical to the MTS criterion [6].

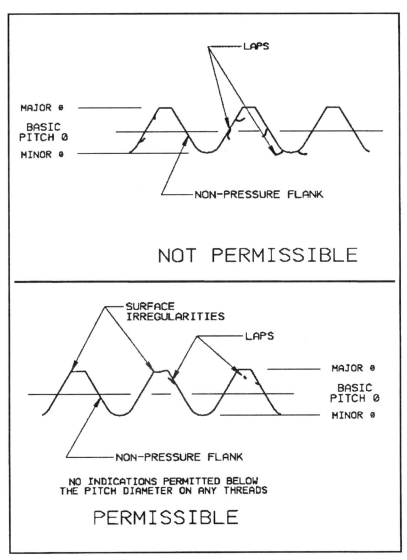

FIG. 1—*Typical thread lap specification call-out (SPS-1-650).*

Trajectory prediction criteria including the MTPS criterion were developed with the assumption of static loading. It has been noted that fatigue crack trajectories differ somewhat from crack trajectories resulting from static loading [2,7]. It is considered though, that various criteria, including MTPS adequately match experimental data for both static and fatigue (cyclic) loading conditions [2,5,7–10]. In some studies the match was actually better for fatigue crack trajectories than for cracks created under static conditions [7]. The proposed inspection criterion includes margins of safety to allow for variations in trajectory.

The MTPS criterion assumes mixed Mode I and Mode II loading, but Mode I crack propagation only. Though Mode II crack propagation under static mixed-mode loading is observed, it is generally found that fatigue cracks propagate in Mode I [9,11,12] even under mixed-mode conditions. Mode I propagation can be complex. Typically a Mode II crack under mixed-mode loading bifurcates, then resumes propagating as a Mode I crack. This is a behavior found by several researchers of mixed-mode fatigue crack propagation [11] and is the general behavior noted by others [2,10].

Mode II fatigue crack growth has been found at near threshold ΔK_I and ΔK_{II} values. These Mode II fatigue cracks propagated short distances (<0.0125 mm) before arresting [11,13]. At increased stress intensities the fatigue crack bifurcated, and only resumed propagating in Mode I. Higher R ratios, which are characteristic of bolt applications, tended to reduce the range of ΔK_I–ΔK_{II} in which Mode II propagation occurred. It is possible that this Mode II propagation was an artifact of the preparation of the test specimens [11]. Margins of safety in the proposed inspection criterion allow for some Mode II propagation.

Continuous Mode II crack propagation has been reported in 75 to 78% ($\varepsilon \approx 2.5$) cold-drawn, highly anisotropic low and high carbon steel wire [14]. The Mode II crack propagation started to appear when the ratio of the applied shear stress to the applied tensile exceeded 0.25 ($K_{II} = 0.42K_I$). It was attributed to the elongated geometry of the microstructure, and did not occur in isotropic wire.

For most aerospace bolt materials, microstructural anisotropy should not be sufficient to substantially effect crack trajectories. A separate large strain FEA study on a high strength-bolting alloy at SPS indicated that strains in threads above the pitch diameter during forming are in the range of 0.80 to 1.3. Only very near the thread flanks might strains exceed 2.0, though no such strains were indicated in the study that used an element size of about 1/25th of the thread thickness. Here thickness is taken in the direction parallel to the thread axis. In this study strains in the center most 80% of the thread thickness did not exceed 1.0. This strain was perpendicular to the bolt axis, and thus perpendicular to the grain orientation in the bolt blank. This straining would thus be expected to reduce anisotropy.

Some aerospace bolts are made from 50% cold-drawn material. In these materials microstructural anisotropy may have an effect on fatigue crack propagation. Special consideration may be needed for thread lap criterion in these materials. Specifically laps with tips near the nonpressure flank may tend to propagate in Mode II, and thus they may not propagate benignly toward the nonpressure flank. For such conditions it may be necessary to stay with current lap inspection criteria.

A set of cyclic load conditions reported to produce continuous Mode II propagation had static stress across the crack face, opening the crack, and cyclic Mode II loading [11]. In this testing the static load was the same order of magnitude as the cyclic load. For thread geometries and application loads, such a combination of static and cyclic loads is unlikely. Bolt preload tends to hold laps or cracks closed in the thread except on the pressure flank near or below the pitch diameter. As will be indicated later, laps in this region would not be allowed by the proposed inspection criterion. Residual stresses in the threads above the pitch diameter would be expected to be low relative to preload induced stresses. This can be understood by noting that, as stated above, all material above the pitch diameter undergoes essentially the same amount of large plastic deformation during thread formation. Variation in the effective stress above the pitch diameter is also small, thus local residual stresses would be expected to be small. For threads rolled after heat treatment, fairly large compressive stresses are induced in the threads. These stresses would tend to hold cracks closed. Threads rolled before heat treatment would be free of residual stresses.

The axisymmetric FEA analysis performed here precludes torsional loading on the bolt. Accordingly Mode III influence on crack propagation is excluded. This omission should not be significant due to the phenomena of torsional relaxation in bolts. It has been found that depending on characteristics of the bolted joint, around 50% of the torsion in a bolt is relaxed when the wrench is removed [15]. One suggestion is that in cyclically loaded bolts, all or nearly all torsional stress disappears.

Procedures and Results

Empirical tests to differentiate potentially harmful lap geometries from benign geometries would be difficult or impossible. Techniques simply do not exist to place laps into threads with sufficient precision for a coherent test program. Empirical tests are complicated by the destruction of the load bearing threads upon bolt failure during laboratory testing. Usually evidence of crack origin is obliterated if it is above the nut minor diameter. Additionally, lap geometries and loading conditions are too complex for closed form analysis.

FEA and fracture mechanics offer a theoretical way to resolve the potential differences in behavior between expected lap geometries. In this study it was assumed that in the worst case thread laps behave like fatigue cracks; fatigue crack propagation behavior was then predicted using FEA and fracture mechanics techniques. Specifically, the maximum tangential principal stress (MTPS) criterion was used to predict fatigue crack trajectory, and the final shape of a fatigue crack.

All FEA was performed with ANSYS versions 5.0, 5.1, and 5.5 running under Windows NT. A two-dimensional axisymmetric model of the full bolt, nut, and joint geometries as seen in Fig. 2 was used. The model consisted mostly of "Plane 42" four-node elements. "Plane 2," six-node triangular solid elements, with the first row of elements around the crack tip singular as recommended by ANSYS, were used to model the crack tip singularity. "Contac 48" contact elements were used to model nut-bolt and nut-joint interfaces such that load (pressure) distribution along the threads was determined automatically. Refer to ANSYS Manuals for exact descriptions of these elements. The contact elements were also used on the crack faces. Crack face coefficients of friction were 0.4 dynamic and 0.56 static. In separate tests thread friction was determined not to have a significant effect and was not used.

To provide confidence in the computer modeling, θ_C from empirical testing of flat 7075-T6 aluminum plates with diagonal cracks from technical literature [10] were compared with results of computer modeling. The results of this work can be found in Table 1. Listed are β, θ_C determined empirically, and θ_C determined using the MTS, MTPS, and SENE criteria, as well as K_I and K_{II}. The initial crack length was 12.7 mm. The FEA data does not differ significantly from the empirical data. All of the criteria predict crack propagation in substantially the same direction.

Steel bolts with an 895 MPa yield strength, tightened to 422 000 N (approximately 90% of the nominal yield strength) were modeled. Both 1.00-18 and 0.375-24 UNJF-3A threads per ASME B1.15 were examined. All bolts modeled had two diameters of shank length below the run out thread. Bolts were mated with typical lightweight aerospace nuts, a geometry that loads the first thread most heavily. Loads were reacted out to a thick plate representing the joint. The cracked thread on each model had elastic-plastic material properties, the rest of the model had elastic properties only.

Loads applied simulated standard aerospace structural fastener fatigue inspection test loads.

Maximum Bolt Load = 52% Bolt Minimum Ultimate Tensile Strength

Minimum Bolt Load = 0.1 × Maximum Bolt Load ($R = 0.1$)

At minimum loads the predicted crack propagation directions were substantially the same

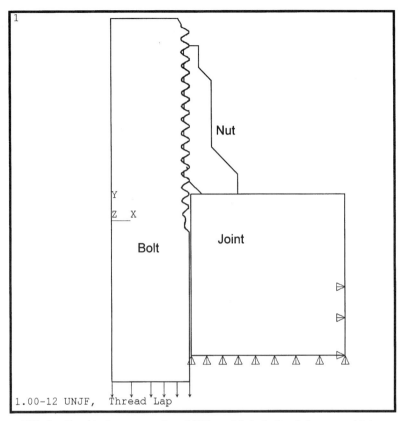

FIG. 2—*Graphical representation of FEA model, including bolt, nut, and joint.*

as those at the maximum loads. Therefore, all propagation directions were determined at the maximum load.

Figure 3 shows FEA predicted maximum principal stress vectors in the first thread in the absence of a lap. The center of the bolt (thread axis) is to the left. The pressure flank of the thread is the bottom edge of the thread, the major diameter is indicated, and the vertical dashed line represents the thread pitch diameter. Note that the maximum principal stresses are aligned with the pressure flank outside the pitch diameter as would be expected for

TABLE 1—*Initial angle of crack extension (θ_C) for diagonal crack in 7075-T6 aluminum plate.*

$\beta,°$	Empirically Determined $\theta_C,°$ [10]	MTS Predicted $\theta_C,°$	MTPS Predicted $\theta_C,°$	SENE Predicted $\theta_C,°$	K_I, MPa in.$^{1/2}$	K_{II}, MPa in.$^{1/2}$
15	81–90	78	76	73	4.1	5.8
30	50–68	64	56	62	8.91	10.2
45	57–62	51	56	50	15.3	11.5
60	45–57	36	34	39	21.6	9.4

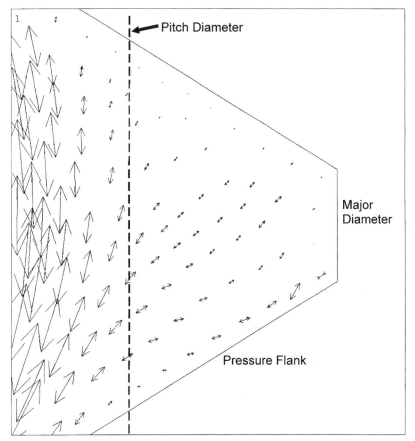

FIG. 3—*Vector plot of maximum principal stress in loaded thread with no cracks.*

cantilever beam type loading. Also note that below the pitch diameter the maximum principal stresses increase in magnitude and become aligned with the bolt axis.

It might seem that a small crack near the pressure flank, propagating according to the MTPS criterion, would start perpendicular to the pressure flank and gradually curve toward the center of the bolt until it got below the pitch diameter. It would then propagate rapidly toward the center of the bolt and eventually fail the bolt. A crack in the thread, though, significantly redistributes stresses in the thread so that determining crack trajectories is much more complicated. Preexisting loads and stresses in the thread can still strongly affect crack propagation. Specifically, the large-magnitude maximum principal stresses below the pitch diameter would be expected to steer a crack entering in the area toward the bolt center, resulting in bolt failure.

In the analysis a single lap was assumed to exist in the first engaged bolt thread nearest the bolt shank. The two-dimensional axisymmetric model restricts this lap to a uniform depth and a 360° extent around the circumference of the bolt. Worst-case geometry for fatigue crack propagation was used. This condition was obtained by using thread flank angles that caused nut-bolt contact near the major diameter, and by using the stiffest nut-and-bolt thread geometry allowable by the specification.

A model without thread laps was run first to establish a baseline of performance. Laps were then grown as fatigue cracks in the following iterative process.

1. A model was built with a crack having the geometry of a thread lap for which the effect on bolt performance was to be determined.
2. The loads were applied, and θ_C was determined using the desired criterion. The MTPS criterion was used except where other criteria were used in comparison.
3. A new model was built with a crack propagated a small distance (0.063 to 0.191 mm) in the direction indicated. Plastic deformations were not accumulated between models. Thus, plastic crack closure behind the crack was not allowed.
4. The loads were then applied to the new model and θ_C determined for the new model.
5. This process was iterated until the crack appeared to be either benign or dangerous.

Figure 4 is a maximum principal stresses contour plot of the first thread, with a lap. The contour legend is along the right-hand side of the plot. It can be seen that the stress distribution is very different from that in Fig. 3. The stresses are highest near the root radius at the bottom of the thread and at the crack tip. A significant feature of the crack propagation can be illustrated with this plot. The piece of thread that was separated from the main body of the thread by the crack was loaded like a cantilever beam. This cantilever beam was fixed to the rest of the thread and bolt along the ligament of material ahead of the crack. Loads

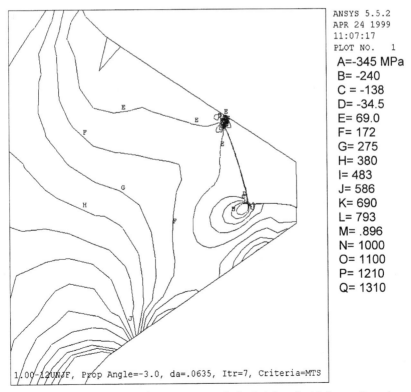

FIG. 4—*Maximum principal stress contours in loaded thread with "grown" crack.*

were applied to the separated piece of thread along the upper surface of the original lap. Such loading is similar to a simple cantilever beam with a uniform load. For such a cantilever beam large stresses are generated in the beam, perpendicular to the direction of loading. Thus the maximum principal stresses in the ligament ahead of the crack are aligned approximately parallel to the nonpressure flank of the thread. With such alignment of the maximum principal stresses, the crack will propagate toward the nonpressure thread flank. This general behavior is true for a family of thread laps.

The effect of using the MTS versus the MTPS criterion, varying the thread flank angle within the allowable range, and of using elastic versus elastic-plastic material were investigated. The lap shown in Fig. 4 was propagated under the four sets of conditions given in Table 2. The resulting four crack trajectories shown in Fig. 5 are nearly identical. For the iterations listed in the first column of Table 2, K_I, K_{II}, and θ_C are listed for each set of conditions. A positive θ_C indicates that the crack turned clockwise toward the nonpressure thread flank. K_I and K_{II} decreased quickly on the first iteration, and decreased more slowly as the crack propagated. K_{II} always decreased well below K_I after the first iteration. As the crack tips approached the nonpressure thread flank, K_I and K_{II} decreased sufficiently that crack self-arrest was probable. Note that on the final iteration of the first and last set of conditions listed, the crack turned sharply toward the nonpressure thread flank. Apparently due to the difficulty of finding a maximum point (as opposed to the zero crossing point for the MTPS criterion) and due to the piecewise continuous nature of finite elements, the MTS criterion-computed θ_C jumped between $+9.0°$ and $-9.0°$. The load carried by the first thread without laps was 49 100 N; with the fully developed lap this load was reduced by only 3.5% to 47 400 N.

The finite-element results obtained in this study appear to be numerically well converged. The conditions listed in the first column of Table 2 were rerun with elements one-half the size, and with one-half the incremental crack extension distance. The results are plotted in Fig. 5. It can be seen that the predicted crack trajectories for each set of conditions are essentially indistinguishable.

Figure 6 is a thread profile with a set of laps and their predicted crack trajectories. Each lap was modeled separately. Note that most of the laps, particularly those originating at the major diameter and the nonpressure flank were predicted to behave quite benignly. Significantly, some laps do not appear as benign. Note the two laps originating from the pressure flank of the thread. The one nearest the major diameter appears at first to head for the nonpressure flank but in the last few iterations turned and headed parallel to the nonpressure flank and toward the thread root. The lap closer to the pitch diameter propagated at 45° to the bolt axis on the first two iterations. But during the last iteration the local maximum principal stresses began to turn the crack toward the center of the bolt. Such a lap should not be permitted.

Superimposed on the thread in Fig. 6 is a polygon representing the proposed inspection criterion. The proposed inspection criterion would permit any lap that has all of its geometry inside the polygon. If any part of the lap is outside the polygon, the lap is nonpermissible. The dimensions of the polygon are given. The zone would extend in from the minimum major diameter toward the bolt center, and down from the nonpressure flank distances of 20% of the truncated thread height. The zone extends to a depth of 10% the truncated thread height along the nonpressure flank below the pitch diameter. Laps on the pressure flank would be allowed only if they originate above the minimum major diameter.

Conclusions

A method of differentiating thread laps that are potentially benign from those which are potentially dangerous was presented. This method is based on fracture mechanics principles

TABLE 2—*Data for cracks grown from the same lap under various conditions.*

Iteration Number	MTPS Criterion, 30° Nut Flank Angle, Elastic Material Properties			MTPS Criterion, 29° Nut Flank Angle, Elastic Material Properties			MTS Criterion, 30° Nut Flank Angle, Elastic Material Properties			MTPS Criterion, 30° Nut Flank Angle, Elastic-Plastic Material Properties		
	K_I, MPa in.$^{1/2}$	K_{II}, MPa in.$^{1/2}$	θ_C, °	K_I, MPa in.$^{1/2}$	K_{II}, MPa in.$^{1/2}$	θ_C, °	K_I, MPa in.$^{1/2}$	K_{II}, MPa in.$^{1/2}$	θ_C, °	K_I, MPa in.$^{1/2}$	K_{II}, MPa in.$^{1/2}$	θ_C, °
1	4.1	7.0	81.3	5.9	9.2	81.4	4.1	7.0	81.0	15.2	42.1	82.7
2	4.8	0.39	-6.3	12.4	1.4	-9.7	4.8	0.36	-9.0	7.1	1.7	-11.1
3	4.9	0.10	-1.7	12.4	0.54	2.4	4.9	0.30	9.0	7.1	0.22	1.1
4	4.6	0.12	-2.1	11.9	0.22	0.23	4.5	0.84	-9.0	6.6	0.18	-2.4
5	4.3	0.11	-1.9	10.7	0.06	-1.0	4.2	0.36	-9.0	6.5	0.27	1.4
6	3.7	0.16	-3.1	8.8	0.09	-2.8	3.7	0.21	9.0	5.0	0.18	-4.5
7	1.9	0.17	-7.3	5.9	0.53	-15.3	...	...	...	4.6	0.09	-1.4
8	...	...	4.3	4.1	0.52	2.2	...	...	...	...	...	2.2
9	0.24	0.14	53.5	...	...	...	...	...	...	2.4	0.43	-15.1
10	...	...	...	...	...	...	...	...	...	0.2	0.45	60.8

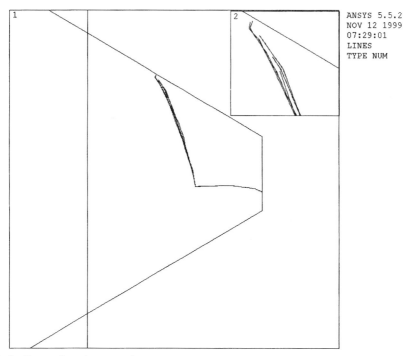

FIG. 5—*Trace of cracks grown from same original lap under four sets of conditions listed in Table 2 and under conditions of Table 2, Column 1, with mesh refinement.*

and finite-element analysis. It represents an improvement over the 30- or 40-year-old intuitive basis for the current thread lap inspection criteria. A concise, less subjective method was presented for inspecting laps.

The proposed inspection criterion is generally more liberal than existing aerospace criterion; that is, more possible laps are allowed. Some laps allowed by current thread lap inspection criteria would, however, be disallowed. End-users are not being asked to allow something new. They already allow bolts with thread laps; this method would simply put a more rational basis behind the acceptance or rejection of thread laps.

Additional work that is needed to better understand potential thread lap behavior includes:

• Studying more lap geometries, including multiple laps and three-dimensional laps.
• Accounting for different materials and microstructural anisotropy.
• Testing.
• Analysis of the effect of erasing plastic deformation between load steps.
• Analysis of the effect of residual stresses.

Note that since not all possible lap geometries within the polygon were considered, laps with significantly different geometries than those modeled should be rejected or in some way checked.

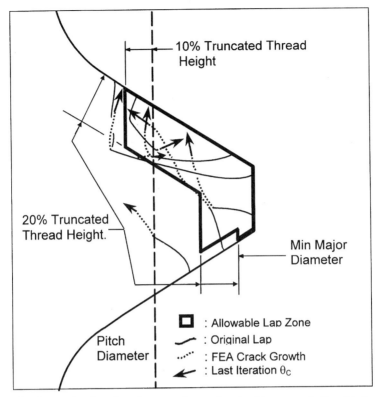

FIG. 6—*Composite plot of cracks grown into a thread, with proposed allowable lap zone.*

Acknowledgments

The author would like to acknowledge Larry Kline of SPS Technologies for initiating reexamination of thread lap inspection criteria behavior at SPS, and for coming up with the Permissible-Zone inspection technique. Also, Dr. Ronald Hartranft, Lehigh University, and Dr. Richard Roberts, Lehigh University.

References

[*1*] Erdogan, F. and Sih, G. C., "On the Crack Extension in Plates Under Plane Loading and Transverse Shear," *Journal of Basic Engineering,* Vol. D 85, 1983, pp. 344–353.
[*2*] Guo, Y. H. and Srivatsan, T. S., "Influence of Mixed-Mode Loading on Fatigue-Crack Propagation," *Engineering Fracture Mechanics,* Vol. 47, No. 6, 1994, pp. 843–866.
[*3*] Hussain, M. A., Pu, M. A., and Underwood, J., "Strain Energy Release Rate for a Crack Under Combined Mode I and Mode II," in *Fracture Analysis, ASTM STP 560,* American Society for Testing and Materials, 1974, pp. 2–28.
[*4*] Theocaris, P. S., Andrianoloulos, N. P., and Kourkoulis, S., "The Angle of Initiation and Propagation of Cracks in Ductile Media," *Experimental Mechanics,* Vol. 27, 1987, pp. 120–125.

[5] Maiti, S. K. and Prasad K. S. R. K., "A Study on the Theories of Unstable Extension for the Prediction of Crack Trajectories," *International Journal Solids Structures,* Vol. 16, 1980, pp. 563–574.

[6] Maiti, S. K. and Smith, R. A., "Criteria for Mixed Mode Brittle Fracture Based on the Pre-Instability Stress-Strain Field-I and II," *International Journal of Fracture,* Vol. 23, 1983, pp. 281–295.

[7] Abdel Mageed, A. M. and Pandey, R. K., "Mixed Mode Crack Growth Under Static and Cyclic Loading in Al-Alloy Sheets," *Engineering Fracture Mechanics,* Vol. 40, No. 2, 1991, pp. 371–385.

[8] Panasyuk, V. V., Zboromriskii, A. I., Ivanitskays, G. S., and Y. Yarema, S., "Applicability of the σ_θ-Criterion for Prediction of a Curved Crack Trajectory," *Strength of Materials,* Vol. 18, No. 9, 1987, pp. 1137–1142.

[9] Qain, J. and Fatemi, A., "Fatigue Crack Growth Under Mixed Mode I and II Loading," *Fatigue Fracture Engineering Material Structure,* Vol. 19, No. 10, 1996, pp. 1277–1284.

[10] Ramulu, M. and Kobayashi, A. S., "Numerical and Experimental Study of Mixed Mode Fatigue Crack Propagation," *Handbook of Fatigue Crack Propagation in Metallic Structures,* Elsevier Science B. V., 1994.

[11] Bold, P. E., Brown, M. W., and Allen, R. J., "A Review of Fatigue Crack Growth in Steels Under Mixed Mode I and II Loading," *Fatigue Fracture Engineering Material Structure,* Vol. 15, No. 10, 1992.

[12] Pook, L. P., "Mixed Mode Fatigue Crack Propagation," *Handbook of Fatigue Crack Propagation in Metallic Structures,* Elsevier Science B. V., 1994.

[13] Hua, G., Alagok, N., Brown, M. W., and Miller, K. J., "Growth of Fatigue Cracks Under Combined Mode I and II Loads," in *Multiaxial Fatigue, ASTM STP 853,* American Society for Testing and Materials, 1985, pp. 184–202.

[14] Verpoest, I., Notohardjono, B. D., and Aernoudt, E., "Fatigue of Steel Wire Under Combined Tensile and Shear Loading Conditions," in *Multiaxial Fatigue, ASTM STP 853,* 1985, pp. 361–378.

[15] Bickford, J. H., *An Introduction to the Design and Behavior of Bolted Joints,* 2nd ed., Marcel Dekker, Inc., 1990, pp. 181–182.

S. R. Mettu,[1] A. U. de Koning,[2] C. J. Lof,[2] L. Schra,[2] J. J. McMahon,[3] and R. G. Forman[3]

Stress Intensity Factor Solutions for Fasteners in NASGRO 3.0

REFERENCE: Mettu, S. R., de Koning, A. U., Lof, C. J., Schra, L., McMahon, J. J., and Forman, R. G., **"Stress Intensity Factor Solutions for Fasteners in NASGRO 3.0,"** *Structural Integrity of Fasteners: Second Volume, ASTM STP 1391,* P. M. Toor, Ed., American Society for Testing and Materials, West Conshohocken, PA, 2000, pp. 133–139.

ABSTRACT; Fasteners are widely used in aerospace and other industrial applications. Structural integrity analysis of such components is a critical necessity in many of the applications. The present paper describes some recently obtained stress intensity factor solutions for cracked threaded members and for cracked fillet areas under bolt heads. The cracks were assumed to be thumb-nail shaped and to originate at the thread roots or fillet radii. The assumed aspect ratios were based on observed shapes from crack growth tests. The tests were conducted on A286 steel, Ti-6Al-4V titanium alloy and 7075-T73 aluminum alloy. Crack growth was measured using marker bands that result from changing the load level after a certain number of cycles. The measured crack growth rates were converted to stress intensity factors using certain principles of similitude. Three-dimensional finite elements were used to verify the solutions. Finite-element analysis was also used to compute the stress concentration factors at the thread roots and fillets. A distinction was made between rolled and cut threads. In the case of rolled threads or fillets, the residual stresses cause a reduction of the stress intensity factor for small cracks. In the case of machined threads or fillets, the stress concentration factor at the thread or fillet root governs the magnitude of the stress intensity factor. Specific solutions were constructed and coded in for standard aerospace bolt sizes, but provision was made for other sizes that may be applicable for general industrial use. The solutions are valid for metric as well as for U.S. customary sizes. These solutions were developed for implementation into the NAS-GRO 3.0 software. NASGRO is a state-of-the-art software package for fatigue crack growth and fracture mechanics analysis.

KEYWORDS: fasteners, bolts, threads, cracks, stress intensity factor, fatigue, finite element

Fasteners are widely used in aerospace and other industrial applications. Structural integrity analysis of such components is a critical necessity in many of the applications. During both the design and service phases, structural integrity assessments typically use the damage tolerance approach. In this approach, flaws that are at the threshold of detection by nondestructive evaluation (NDE) methods are assumed to be present in the structure at the worst possible location. The residual strength of the structure in the presence of the assumed flaw should be adequate for safe operation. In addition, the propagation of such a flaw subject to service loads until it reaches a size that leads to catastrophic failure needs to be characterized. Central to such characterization is the knowledge of the stress intensity factor for the assumed

---

[1] Advanced Systems Engineering specialist, Lockheed Martin Space Operations, Houston, TX 77058.
[2] Research engineers, National Aerospace Laboratory (NLR), Emmeloord, The Netherlands.
[3] Materials engineer and senior scientist, respectively, NASA Johnson Space Center, Houston, TX 77058.

crack geometry. The present paper is concerned with constructing stress intensity factor solutions for cracked threaded members and for cracked fillet areas under bolt heads. These solutions were developed for implementation into the NASGRO 3.0 software. NASGRO is a state-of-the-art software package for fatigue crack growth and fracture mechanics analysis. A brief review of the available solutions for threaded fasteners is first presented. Subsequent sections deal with the methods used to develop the solutions. Finally, the engineering models based on these solutions are presented.

Literature Survey

Early studies [1–3] of cracks in solid cylinders form the basis for estimating stress intensity factors for cracks in the shank area of bolts. Experimental verification of these solutions was conducted by Forman and Mettu [4]. The stress intensity factor solution for a crack in the threaded area is a more complicated problem and few good solutions exist. The fillet area under the bolt head can be treated similar to the threaded area by using suitable stress concentration factors. Liu [5] presented a comprehensive review of the available solutions for threaded bolts. In his review, he compared the available analytical solutions with experimental results for tension and bending loads. None of these solutions takes the interaction between the nut and the bolt into account. The present study aims to represent the loading more accurately by considering the bolt/nut assembly for modeling purposes in both the analytical and experimental approaches.

Experimental Method

The experimental work described in this paper was conducted by de Koning, Lof and Schra [6]. In order to simulate realistic loading conditions, threaded bolt/nut assemblies loaded in tension and bending were used. Aerospace-quality bolts M8*1.00 mm and M12*1.25 mm made of three different materials, A286 steel, Ti-6A1-4V titanium alloy and 7075-T73 aluminum alloy, were used. The nut was made of steel. Constant-amplitude loading was applied. To measure the crack growth, marker bands were induced by reducing the load level serially at predetermined numbers of cycles. While the constant-amplitude loading was applied at a stress ratio of $R = 0.1$, the marker loads were applied at a ratio of $R = 0.7$. The mean load level was kept constant. During some of the tests, the level of bending stresses in the bolt shank was measured using strain gages. These measurements were made at the beginning of the test, both before the fatigue crack was initiated and during the crack growth phase, to ensure that the bending stresses were small. The maximum loads applied for these tests were 23.5 kN and 9.7 kN for the M12 and M8 bolts, respectively. The elliptical starter notch was prepared using electric discharge machining with a depth of 0.5 mm and an a/c ratio of 0.6.

The stress intensity factor (K) along a crack front in the bolt is determined using a principle of similitude. According to this principle, it is assumed that for identical environmental conditions, the stress intensity factor is the same if the fatigue crack growth rate is the same, regardless of geometry and loading conditions. Using the marker bands mentioned above, the crack size can be measured at regular intervals from the fracture surface. Knowing the corresponding cycle count, the crack growth rate versus crack size can be plotted. Using a cylindrical bar specimen, a curve of crack growth rate versus stress intensity factor range $(dc/dN$ vs. $\Delta K)$ can be generated, because the stress intensity factor solution is known for this specimen [7]. Using these two plots, the K versus c relation for the cracked bolt can be deduced as shown in Fig. 1. The same procedure is repeated at various points along the crack front.

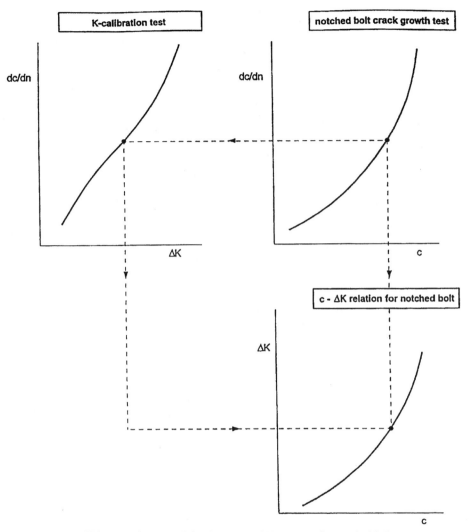

FIG. 1—*Schematic of the derivation of* K *versus* c *for cracked bolt.*

The fatigue tests on the notched bolts with cut threads showed quite a different behavior compared to the bolts with rolled threads. For bolts with rolled threads, the crack aspect ratio remained at about $a/c = 1$ and for machined fillets, the crack aspect ratio is found to be close to $a/c = 0.645$. A detailed account of the behavior of various types of starter notches from cut and rolled threads is given in Ref 6. Based on observations, only some specimens were selected for determination of the stress intensity factors.

Finite-Element Analysis

A portion of the finite-element analysis described here was first reported by de Koning et al. in Ref 6 and the rest was conducted recently. The load transfer between the threaded bolt and the nut was analyzed using the finite-element method. This method was also used to compute the stress intensity factor distribution along the crack front. Full three-dimensional meshes of nut and bolt were used in the analysis. Quadratic isoparametric 20-noded brick elements were used. Elements adjacent to the crack front were shrunk to form wedge-shaped elements to obtain the square root variation of displacements near the crack tip. The commonly used shifting of mid-side nodes to the quarter-point position was used. The stress intensity factors were extracted using de Lorenzi's method [8]. This method uses the virtual crack extension to first compute the energy release rate and then the stress intensity factor assuming plane strain conditions. The discretized bolt length was seven times the pitch. Four threads under the nut and three threads on the free shank portion of the bolt were chosen for modeling. A typical finite-element mesh had about 7000 elements. The contact surface of the nut and bolt is simulated by rigid coupling of displacements of all nodes on the contact surface. This implies that friction is not modeled. The outer surface of the cylinder is modeled as a cylinder rather than as a hexagonal pyramid.

The p-version finite-element software "STRESS CHECK" was used to compute the stress concentration factors (K_t) at the root of a fillet. The stress concentration factors for a given fillet radius to shank diameter ratio, r/D, is shown in Table 1.

Implementation in NASGRO

The following three crack configurations have been formulated for engineering use and implemented into NASGRO 3.0 as described in the reference manual [9].

SC08—Surface Crack in Bolt Thread (Thumbnail Crack)

Figure 2 shows the geometry and loading. Stresses S_0 and S_1 are the remote applied stresses. Table 2 lists values of normalized stress intensity factors F_0 and F_1. These factors are defined as $F_0 = K_1/(S_0\sqrt{\pi a})$ and $F_1 = K_1/(S_1\sqrt{\pi a})$. The parameter f_x is defined as

$$f_x = [1 + 1.464(a/c)^{1.65}]^{-1/2}$$

SC13—Surface Crack from Cut Fillet Under a Shear Bolt Head

Figure 3 shows the crack geometry. Values of the normalized stress intensity factors F_0 and F_1 from Table 2 and the stress concentration factors from Table 1 are used.

SC14—Surface Crack from Rolled Fillet Under a Tension Bolt Head

Figure 4 shows the crack geometry. Values of F_0 and F_1 from Table 2 are used for this crack case. For the machined fillet, $a/c = 0.645$ is used and for the rolled fillet, $a/c = 1.0$ is used. In this case, the fillet radius is assumed constant ($r/D = 0.1$).

TABLE 1—*Stress concentration factors at fillet for a bolt in tension and bending.*

r/D	0.005	0.01	0.015	0.02	0.025	0.03	0.035	0.04	0.045	0.05
K_t	10.8	7.89	6.55	5.73	5.17	4.77	4.48	4.19	3.97	3.79
r/D	0.055	0.06	0.065	0.07	0.075	0.08	0.085	0.09	0.095	0.10
K_t	3.63	3.49	3.37	3.26	3.16	3.07	2.97	2.91	2.84	2.78

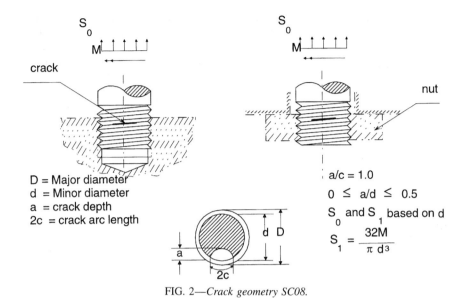

FIG. 2—*Crack geometry SC08.*

Summary

Stress intensity factor solutions were developed for cracked threaded members and fillet areas under bolt heads. The cracks were assumed to be thumbnail shaped (with suitable aspect ratios), originating at the thread roots or fillet radii. The assumed aspect ratios were based on observed shapes from crack growth tests conducted on A286 steel, Ti-6A1-4V titanium alloy and 7075-T73 aluminum alloy. The measured crack growth rates were con-

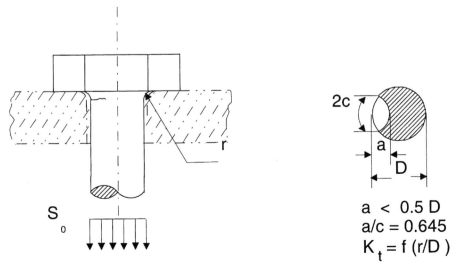

FIG. 3—*Crack geometry SC13: shear or machine bolt—machined fillet.*

TABLE 2—*Stress intensity factors for a bolt in tension and bending.*

	F_0		F_1	
a/D	$a/c = 0.645$	$a/c = 1.0$	$a/c = 0.645$	$a/c = 1.0$
0.0	$K_t f_x$	1.00	$K_t f_x$	0.60
0.05	...	0.84	...	0.54
0.1	0.95	0.76	0.61	0.48
0.2	0.90	0.65	0.54	0.37
0.3	0.98	0.59	0.55	0.31
0.4	1.29	0.62	0.64	0.30
0.5	2.05	1.0	0.84	0.50

verted to stress intensity factors using principles of similitude. Three-dimensional finite elements were used to verify the solutions. Finite-element analysis was also used to compute the stress concentration factors at the thread roots and fillets. A distinction was made between rolled and cut threads. In the case of machined threads or fillets, the stress concentration factor at the root governs the magnitude of the stress intensity factor. Specific solutions suitable for engineering use were constructed and coded in for standard aerospace bolt sizes into the NASGRO software. Provision was also made for other sizes that are applicable for general industrial use.

Acknowledgments

The authors from NLR acknowledge the financial support provided by the European Space Agency during the course of this research, and the author from Lockheed Martin acknowledges the support provided by the National Aeronautics and Space Administration under contract NAS9-19100. The first author also appreciates the helpful comments by Dr. V. Shivakumar and the rendering of figures by Mr. Feng Yeh, both of Lockheed Martin.

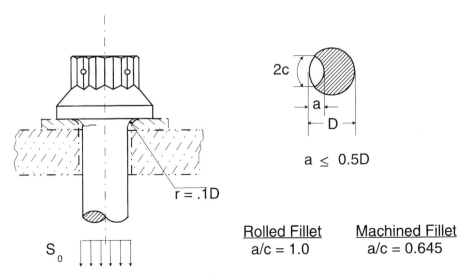

FIG. 4—*Crack geometry SC14: tension bolt—crack in bolt head fillet.*

References

[1] Tada, H., Paris, P. C., and Irwin, G. R., *The Stress Analysis of Cracks Handbook,* Del Research Corporation, Hellertown, PA, 1973.

[2] Benthem, J. P. and Koiter, W. T., "Asymptotic Approximations to Crack Problems," in *Methods of Analysis and Solutions of Crack Problems,* G. C. Sih, Ed., Noordhoff International Publishing, Groningen, 1973, pp. 131–178.

[3] Forman, R. G. and Shivakumar, V., "Growth Behavior of Surface Cracks in the Circumferential Plane of Solid and Hollow Cylinders," *Fracture Mechanics: Seventeenth Volume, ASTM STP 905,* J. H. Underwood, R. Chait, C. W. Smith, D. P. Wilhelm, W. A. Andrews, and J. C. Newman, Eds., American Society for Testing and Materials, West Conshohocken, PA, 1986, pp. 59–74.

[4] Forman, R. G. and Mettu, S. R., "Behavior of Surface and Corner Cracks Subjected to Tensile and Bending Loads in Ti-6A1-4V Alloy," in *Fracture Mechanics: Twenty-second Symposium, Volume 1, ASTM STP 1131,* H. A. Earnst, A. Saxena, and D. L. McDowell, Eds., American Society for Testing and Materials, West Conshohocken, PA, 1992, pp. 519–546.

[5] Liu, A. F., "Evaluation of Current Analytical Methods for Crack Growth in a Bolt," in *Durability and Structural Reliability of Airframes—ICAF 17,* Vol. 2, Engineering Materials Advisory Services Ltd., 1993, pp. 1141–1155.

[6] de Koning, A. U., Lof, C. J., and Schra, L., "Assessment of 3D Stress Intensity Factor Distributions for Nut Supported Threaded Rods and Bolt/Nut Assemblies," NLR CR 96692 L, National Aerospace Laboratory (NLR), 1996.

[7] Raju, I. S. and Newman, J. C., "Stress Intensity Factors for Circumferential Surface Cracks in Pipes and Rods under Tension and Bending Loads," NASA TM 87594, August 1985.

[8] de Lorenzi, H. G., "On the Energy Release Rate and the J-Integral for 3-D Crack Configurations," *International Journal of Fracture,* Vol. 19, 1982, pp. 183–193.

[9] *Fatigue Crack Growth Computer Program NASGRO Version 3.00,* JSC-22267B, NASA Johnson Space Center, Houston, TX, 1999.

Testing Procedures

Alan R. Kephart[1]

Fatigue Acceptance Test Limit Criterion for Larger-Diameter Rolled Thread Fasteners

REFERENCE: Kephart, A. R., **"Fatigue Acceptance Test Limit Criterion for Larger-Diameter Rolled Thread Fasteners,"** *Structural Integrity of Fasteners: Second Volume, ASTM STP 1391*, P. M. Toor, Ed., American Society for Testing and Materials, West Conshohocken, PA, 2000, pp. 143–161.

ABSTRACT: This paper describes a fatigue lifetime acceptance test criterion by which studs having rolled threads, larger than 1.0 in. (25 mm) in diameter, can be assured to meet minimum quality attributes associated with a controlled rolling process.

This criterion is derived from a stress-dependent, room temperature air fatigue database for test studs having 0.625 in. (16 mm) diameter threads of Alloys X-750 HTH and direct aged 625. Anticipated fatigue lives of larger threads are based on thread root elastic stress concentration factors which increase with increasing thread diameters. Over the thread size range of interest, a 30% increase in notch stress is equivalent to a factor of five (5×) reduction in fatigue life. The resulting diameter-dependent fatigue acceptance criterion is normalized to the aerospace rolled thread acceptance standards for a 1.0 in. (25 mm) diameter, 0.125 in. (about 3 mm) pitch, Unified National thread with a controlled root radius (UNR). Testing was conducted at a stress of 50% of the minimum specified material ultimate strength, 80 ksi, (552 MPa) and at a stress ratio (*R*) of 0.10. Limited test data for fastener diameters of 1.00 to 2.25 in. (25 to 60 mm) are compared with the acceptance criterion.

Sensitivity of fatigue life of threads to test nut geometry variables was also shown to be dependent on notch stress conditions. Bearing surface concavity of the compression nuts and thread flank contact mismatch conditions can significantly affect fastener fatigue life. Without improved controls these conditions could potentially provide misleading acceptance data. Alternative test nut geometry features are described and implemented in the rolled thread stud specification, MIL-DTL-24789(SH), to mitigate the potential effects on fatigue acceptance data.

KEYWORDS: fasteners, fatigue acceptance tests, rolled threads, thread size effects, Alloy X-750 HTH, Alloy 625 (direct aged), test nut effects, fatigue test procedures, MIL-STD-1312-11A, MIL-DTL-24789(SH)

The rolled thread process has been shown to be an effective means of improving the stress corrosion cracking (SCC) initiation resistance of threaded fasteners [*1–3*]. Rolled threads have an extensively documented improved fatigue resistance over cut threads, particularly for aerospace, petrochemical and transportation industry applications. As a means of supporting improved SCC resistance, fatigue acceptance testing of rolled thread studs is performed to demonstrate the presence of beneficial residual compressive stresses in the rolled thread roots.

Initially the specification for rolled thread studs contained process-related quality assurance requirements consistent with the 40-year aerospace rolled thread experience, typified by aerospace fastener specifications for IN-718 bolts [*4,5*]. Test procedures complied with MIL-STD-1312, Test 11, tension fatigue [*6*]. Use of the aerospace acceptance practice and AMS/

[1] Senior materials engineer, Lockheed Martin Co., Schenectady, NY 12301-1072.

SAE experience would assure that the inherent benefits of the thread rolling process to fastener properties would be linked to commercial process quality controls.

A significant finding of the fatigue acceptance testing of in-feed rolled threads is the occurrence of fatigue fracture in large-diameter threads (2.0 to 2.25 in. basic thread diameter, or 50 to 60 mm) at cycles well below the aerospace minimum acceptance criteria of 65 000 cycles, average. Also, large-diameter cut threads exhibited reduction in fatigue life compared to smaller diameter threads.

Review of these events led to the conclusion that large-diameter threads with more severe notch stresses, when tested at the same net-section stress (50% of ultimate tensile strength, UTS), were responsible for the reduced lifetime, rather than a reduced fatigue benefit due to the thread rolling process. This contention is consistent with fundamental notched fatigue behavior in which the principal driving force of fatigue life is the local notch root stress, and not the net-section stress which is independent of notch sharpness.

To assure consistency of thread root applied notch stress effects, two approaches were considered. In one, the nominal (net-section) test stress was decreased to account for the higher notch concentration factor present in larger-diameter threads. This would sustain the same thread root stress and provide a controlled measure of thread rolling process consistency. The other approach was to sustain the same nominal stress, as in smaller aerospace fasteners, but reduce the fatigue acceptance lifetimes for larger diameter high notch stress threads.

Accordingly, MIL-DTL-24789(SH) [7] contains an allowance for reduced fatigue life when the thread diameter exceeds 1.00 in. (25 mm). This paper describes a procedure for determining the reduced fatigue cycle lifetime for large-diameter threads. The procedure is based on a combination of fundamental notch fatigue behavior and stress-dependent test data from laboratory scale in-feed rolled threads of alloys X-750 HTH and direct aged 625 (DH).

Aerospace Rolled Thread Fatigue Acceptance Testing

Aerospace fasteners and tension fatigue testing include, depending on the specific fastener alloy, the following requirements or features:

(1) Test procedures that comply with MIL-STD-1312, Method 11A (tension fatigue) which describes the compression nuts and facility/fixture requirements for thread testing.

(2) A ratio of minimum to maximum tension test stress (R ratio) of 0.10.

(3) A test load that is normalized to thread section area providing the same net-section test stress for all sized fasteners, independent of thread diameter or thread root radius.

(4) Tests are conducted in the load controlled mode.

(5) Nearly all fastener alloys must meet a minimum fatigue lifetime of 65 000 cycles based on the average of all fatigue tests.

(6) Nearly all fastener alloys must meet a minimum fatigue lifetime of 45 000 cycles based on the minimum individual fatigue test result.

(7) Tests are terminated when fatigue exposure exceeds 130 000 cycles without fracture.

(8) The test result is invalid if the threads fail in shear.

(9) A typical test stress is 50% of the minimum allowed ultimate tensile strength (UTS) of the fastener alloy; however, some special high-strength alloys allow up to 60% UTS test stress for extreme duty service and some other alloys allow lower percentages of UTS because of their limited ability to meet the minimum fatigue cycle lifetime of 65 000 cycles when tested at 50% UTS.

(10) Use of fine-pitch UNJ thread form with a larger root radius than UNR for most applications.

(11) Most aerospace fasteners are bolts having thread diameters of about 0.5 in. (12 mm) or less with infrequent maximum diameters up to 1.5 in. (40 mm).

Evidence of Fastener Size Effects on Fatigue Life

In addition to the observations of apparent thread size effects on fatigue life from experience with rolled and machined (sharp tool cut) fasteners, summarized in Table 1, size effects have also been reported from aerospace fastener experience [8]. These test data from SPS Technologies are displayed in Fig. 1 for high-strength low-alloy steel fasteners of sizes ranging from 0.25 to 5.0 in. (6 to 120 mm) in diameter.

This report shows that large-diameter threads have much lower fatigue cycle lifetimes. Interpretation of the cause of the findings by the author of Ref 8 focuses on microstructure and more likely weak-link materials inclusion related degradations (Weibull sources) for larger-diameter fasteners, without reference to notch effects on thread root stresses.

A review of fastener "size" effect findings and theories shows inconsistent views. A more complete list of the possible fatigue life influencing factors is provided below for controlled test stress conditions; i.e., the same normalized test load (net-section stress) based on test load divided by the thread section area, and fatigue test data determined by thread section fatigue fracture.

- Notch sharpness or bluntness: normally expressed as a stress or strain concentration factor and typically elastic for simplicity.
- Thread fabrication method: surface effects due to altered material properties or manufacturing induced residual stresses or surface roughness when residual stresses are controlled to similar levels.
- Fatigue crack growth path: increases with large size; therefore, implying longer life for larger threads in contrast to the observed shorter life.
- Statistical effects of material inclusions: can act as local stress risers to reduce lifetimes in larger threads with more critical surface exposed, and more likely inclusion content or influence.

Referring to the more fundamental structural analysis work of Peterson and Neuber, it is clear that the principal factor affecting the survival of a notched and tensile stressed component is the influence of notch geometry on local notch strain or stress. Since both the thread root radius and diameter of constant pitch threads mutually affect the fundamental notch stress effects, they cannot be treated independently. Therefore, fatigue acceptance criteria for rolled threads should address this notch stress factor.

Using elastic finite-element notch stress analysis consistent with notch size scaling concepts, the SPS data of Fig. 1 were reevaluated to determine if the reported size effect would remain after appropriate notch factors were included. The SPS reported thread section stress was taken at a fatigue fracture lifetime of 100 000 cycles which is similar to the aerospace acceptance limit. Figure 2 shows that incorporation of notch stress effects, using the simple "linear elastic" rule, would completely compensate for the apparent thread size effect since there is no remaining notch stress trend for either cut or rolled threads over the size range reported.

Numerical Representation of Thread Notch Stresses

Finite-element analysis provided an estimate for notch stress effects which increase with diameter when the thread pitch and notch radius are held constant. The concept of scaling

TABLE 1—*Fatigue fracture lives for cut and in-feed rolled threads of large diameter studs and smaller baseline threads of Alloys X-750 and aged 625 and tested at 75°F and R = 0.1.*

Thread Size and Form, Inches	Material Alloy	Manufacturing Process	Stress Max, Ksi	Root Radius mils	Stress Conc. Factor	Notch Stress Amplitu. SES, Ksi	Fastener Fatigue Fracture Life, cycles	Fatigue Runouts* No. / Cycles
1.000-8UNR-2A	X-750 HTH	Cut	82.4	15	6.3	234	13,396; 14,638; 15,073; 14,730	0
1.375-8UNR-2A	X-750 HTH	Cut	82.6	14.5	7.2	268	13,293; 18,800; 16,800; 13,728; 15,323; 13,109	0
2.000-8UNJ-2A	X-750 HTH	Cut	80.1	21	7.35	265	9,349; 8,490	0
2.000-8UNJ-2A	625 DH	Cut	80.1	21	7.35	265	5,276; 5,804	0
2.250-8UNJ-2A	625 DH	Cut	80.1	21	7.50	273	7,534	0
0.625-11UNR-2A	X-750 HTH	Roll-REC	76.5	11.5	5.7	196	114,000; 140,000; 145,000; 339,000 (Test by SPS)	1 / 260,000
0.625-11UNR-2A	X-750 HTH	Roll-NP	83	11.5	5.7	214	192,475	1 / 200,000
0.625-11UNR-2A	X-750 HTH	Roll-NP	95	11.5	5.7	245	51,662; 54,308; 54,616; 67,878; 70,504; 82,212; 89,925	0
0.625-11UNR-2A	625 DH	Roll-UF	90	11.5	5.7	231	109,211; 111,331; 116,922; 132,685; 173,262; 175,944	2 / 260,000
1.000-8UNR-2A	X-750 HTH	Roll-UF	82.4	15	6.3	234	89,187; 84,215	13 / 200,000
1.375-8UNR-2A	X-750 HTH	Roll-UF	82.5	16	7.0	260	65,247; 67,838; 75,385; 85,045; 94,379; 99,857 100,719; 115,304; 116,919; 121,552; 125,657	4 / 200,000
1.375-8UNR-2A	X-750 HTH	Roll-UF	82.5	17	6.85	253	73,015; 74,946; 72,637; 69,116; 62,691; 54,543; 57,632; 70,427	0
1.375-8UNR-2A	X-750 HTH	Roll-UF	69.5	16	7.0	219	None	7 / 200,000
1.6875-8UNR-2A	X-750 HTH	Roll-NP	80.0	16	7.35	262	49,000; 50,000; 66,000; 74,000; 98,000; 102,000	0
2.000-8UNR-2A	625 DH	Roll-UF	80.6	19	7.5	272	22,116; 23,162; 40,675	0
2.000-8UNR-2A	625 DH	Roll-UF	60.4	19	7.7	204	175,055	1 / 260,000
2.000-8UNJ-3A	X-750 HTH	Roll-NP	80.3	21	7.35	264	39,281; 47,374; 39,117	0
2.000-8UNJ-3A	625 DH	Roll-NP	80.3	21	7.35	264	39,341; 38,807; 36,415; 44,412; 39,262; 40,144; 30,268; 26,042	0
2.000-8UNJ-3A	625 DH	Roll-NP	67.7	21	7.35	222	90,122	0
2.250-8UNJ-3A	625 DH	Roll-NP	81	19.5	7.60	277	23,089; 27,075; 27,161; 27,513; 28,090; 31,318	0
2.250-8UNJ-3A	625 DH	Roll-NP	66	19.5	7.60	226	89,029	0

NOTES: UF = Underfilled Thread Crests;   NP = Nearly Packed Thread Crests due to slightly oversized blank diameters;   Runouts are without thread section fracture

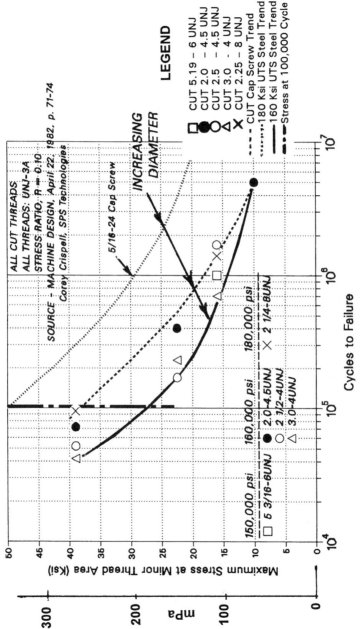

FIG. 1a—Air fatigue test results for cut threads of high strength alloy steel as a function of thread diameter.

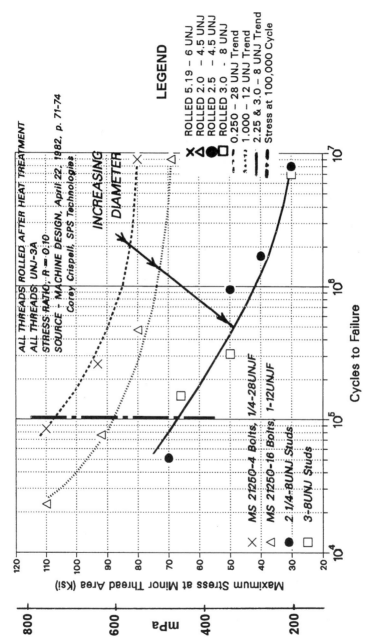

FIG. 1b—Air fatigue test results for high strength alloy steel thread rolled after heat treatment as a function of thread diameter.

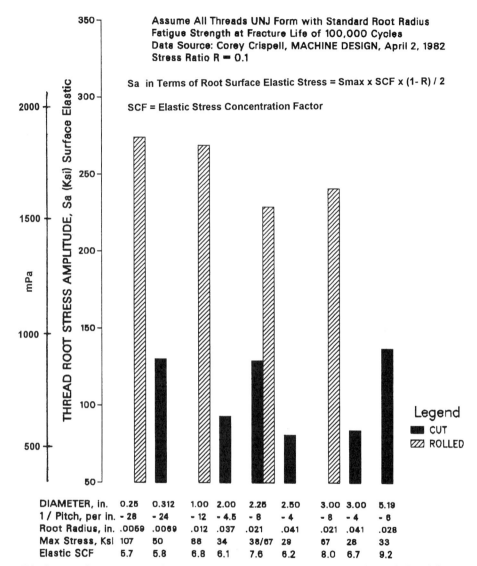

FIG. 2—*SPS fatigue strength for high strength alloy steel of both cut and rolled thread fasteners revised to account for thread root surface elastic stress.*

provides a simple visualization of this size effect. For example, for 8 threads per inch (0.125 in. pitch or about 3 mm), the elastic stress concentration factor (SCF) for the first nut engaged the 1.375-8 UNR-2A thread (M35×3) with a 15 mil root radius (0.4 mm) is 7.0; whereas, doubling the diameter to 2.75 in. (70 mm) results in a SCF of 8.5 for the same thread pitch and root radius. However, if the 1.375 (M35) diameter thread were scaled up by a factor of 2×, the pitch and thread root radius would also increase by 2× to 0.25 in. and 30 mils (6 mm and 0.8 mm), respectively. The resulting SCF for this 2.75-4UNR (M70×6) thread would remain the same (7.0) as for the 1.275-8 UNR (M35×3) thread.

SCF calculated values represent the average elastic stress concentration in the first compression nut engaged thread. Here the SCF is largest since succeeding threads have smaller SCF's due to decreasing load sharing distribution [9]. While these notch stress factors are based on elastic modeling, the resulting effects are also related to notch strain which is considered to be the principal influence on fatigue life.

It is also noted that the coarse thread series with variable pitch has a smaller sensitivity of SCF to increasing size; thus, larger thread diameters in the coarse series would be expected to reveal longer fatigue lives. Likewise, use of the UNJ larger thread root radius series (25% larger than the UNR series) would have a slightly lower notch stress (8%) with a resulting longer fatigue (about 2×) life for the same thread pitch.

Experience with Fatigue Tests of Large-Diameter Threads

All of the large thread fatigue test results have been obtained from Alloy X-750 HTH or 625 DH fasteners. Figure 3 shows the decreasing fatigue life trend with increasing thread diameter for both cut and rolled (in-feed) thread fasteners tested at maximum thread section stresses of about 50% UTS, or 80 ksi (552 MPa). Table 1 lists the existing large thread data and associated fastener features.

While it can be conjectured that the reduced life of the large-diameter rolled threads is due to degraded rolling process and controls, the similar trend from controlled machining process (sharp tool cut) threads implies that the rolling process quality is not the principal cause of the reduced fatigue life.

Reduced Life Acceptance Fatigue Test Criteria

Since nearly all of the large diameter thread acceptance tests were performed only at the required test stress of about 50% of the UTS of the fastener material, direct measurement of a stress effect on fatigue life was not available for large threads. Thus, data from the smaller diameter laboratory stud threads (0.625-11 UNR-2A or M16×2) were used to determine the local thread root stress effect. Table 2 and Fig. 4 list the fatigue fracture data and illustrate stress dependency trends, respectively. Figure 4 shows that, for the stress ratio R of 0.10, the data trend is linear over the range of 40 to 70% of the UTS. Thus, a simple linear numerical representation is facilitated.

Statistical analysis shows excellent reproducibility (small data variability) where the lower one standard deviation occurs at 75% of the mean data. This small data scatter is well within the often quoted factor of two (2×) data range, or a lower life of 50% of the mean data.

Figure 4 displays the data in the standard S-N form (logarithm of stress amplitude (Sa) versus number of cycles to fracture). However, to address thread root notch stresses, the Sa is reflected in terms of surface elastic stress (SES) by multiplying the section stress by the elastic stress concentration factor of 5.7 for the 0.625-11 UNR-2A (M16×2) thread with a root radius of 11.5 mils (0.3 mm). This is consistent with the linear notch stress rule. Statistical analysis of the data of Table 2 and Fig. 4 showed no significant difference between the two fastener alloys for either cut or rolled threads.

The log-linear stress amplitude versus fatigue fracture life trend for the same stress ratio ($R = 0.10$) is of the form:

$$Sa \ (SES, \ ksi) = A \ (N, \ cycle \ life)^n \qquad (1)$$

where A and n are fitting constants.

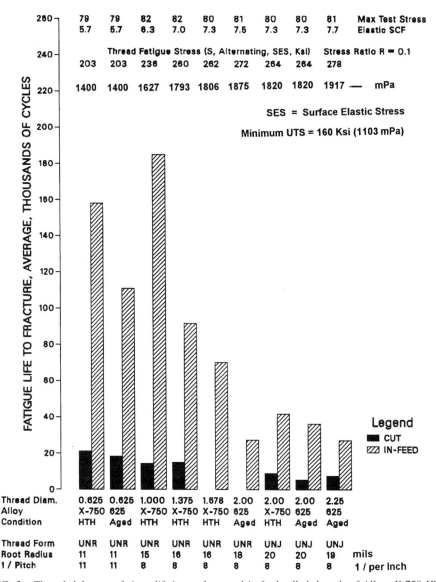

FIG. 3—*Threaded fastener fatigue lifetimes of cut and in-feed rolled threads of Alloys X-750 HTH and 625 DH stressed to 50% UTS as a function of thread diameter.*

Statistical analysis of these data resulted in fitting constant values (A/n) of 1395/-0.1606 for the average life and 1293/-0.1606 for the 95% lower bound. The predicted average and 95% lower-bound lives were 148 313 and 92 492, respectively, for a surface elastic stress amplitude of 206 ksi (1420 MPa).

Based on this stress dependency and increased thread root stress for larger diameter threads, the equivalent fatigue lifetimes for larger-diameter threads can be estimated. Table

TABLE 2—*Fatigue fracture lives for cut and in-feed rolled threads (0.625-11 UNR-2A) of Alloys X-750 and aged 625 tested at 75°F and R = 0.1.*

TEST LOAD Kips Max.	STRESS % of MIN UTS	MAX TEST STRESS Ksi	ALTER-NATING NOTCH STRESS Ksi	FASTENER FRACTURE FATIGUE LIFETIME, Cycles			
				ALLOY X-750 HTH		DIRECT AGED ALLOY 625 DH	
				CUT THREADS	ROLLED THREADS	CUT THREADS	ROLLED THREADS
34.5	98	157	406	750	1,321	1,202	2,088
29.9	85	136	352	2,370	4,796	2,360	5,186
26.4	75	120	310	3,841 3,860	11,055 9,423	3,994	13,189
22.9	65	104	269	7,524	22,455		35,318
21.1	60	96	248	7,796	35,617		
19.4	55	88	228	13,791	64,329	10,937	128,997
17.6	50	80	207	21,234	157,984	20,222 18,405	105,807 116,399
14.1	40	64	166	51,251 68,016	1,485,860 730,577	39,358	807,331
12.3	35	56	145		2,840,764		
10.6	30	48	124	165,508	6,190,642		10,213,625
8.5	24	39	100	803,527			
7.5	21	34	88			1,177,513	

NOTES: 1) For solid 0.625" (16 mm) diameter studs, the stress concentration factor (K$_t$) or SCF = 5.73 Threads have Nearly Packed Thread Crests due to supplier selection of oversized blank diameters

3 lists the features of the larger threads and expected reduced fatigue life, assuming that the "size" is dominated by thread root stress determined by root radius and thread diameter based elastic notch stress concentration factors.

Thread sizes of Table 3 cover the range of testing experience obtained from large (2.0 in. (50 mm) diameter) studs in addition to the laboratory test studs of 0.625 in. (16 mm) diameter. These estimated lives for large-diameter rolled threads are derived directly from the testing of nearly packed in-feed rolled 0.625 in. (16 mm) diameter threads.

Figure 4 also shows the higher stress amplitudes calculated for the larger diameter threads, all of which are based on thread root surface elastic stresses listed in Table 3.

Using the 95% lower bound life of 58 058 for the 1.00 in. (25 mm) thread normalized to the aerospace acceptance of 65 000 cycles, the thread size dependent fatigue life limits are as shown in Fig. 5 for 8-UNR rolled threads. Note that the size-lifetime trends are different for the constant pitch series (8 threads per inch, about 3 mm pitch) and the variable pitch (coarse) thread series because of the difference in thread root radii. Also, acceptance limits for the 8-UNJ threads would be larger (100 000 cycles for the 1.0 in. (25 mm) threads), because of the larger root radii (1.25 times that of 8 UNR) and lower notch root surface elastic stress.

The minimum individual fatigue test acceptance life for the 1.00-8 UNR (M25×3) thread of 40 730 cycles is normalized to the standard aerospace lifetime of 45 000 cycles. From the log-normal distribution statistics for the 0.625 in. (16 mm) diameter thread database, the 40 730 cycle life is equivalent to the 99.8% lower bound, or about three standard deviations

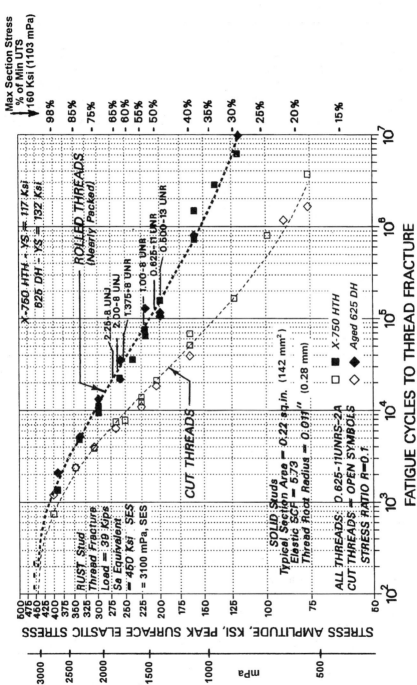

FIG.4—*Threaded fastener fatigue lifetime graph of Table 2 for cut and in-feed rolled 0.625 in. diameter threads of Alloys X-750 HTH and 625 DH, stressed from 30 to 98% of the minimum acceptable UTS.*

TABLE 3—*Estimated fatigue lifetimes of rolled thread fasteners based on thread root stress effects.*

Thread Diameter, Form for 2A Class Thread Fit	Average Root Radius, mils	Elastic SCF	Sa (SES) for 50% of minimum UTS	ESTIMATED FATIGUE LIFETIME Cycles to Fracture		
				Mean Life	95% Low Bounds	99.8% Low Bounds
0.625 - 11 UNR	11.4	5.73	206	148,313	92,492	64,885
1.000 - 8 UNR	15.7	6.16	222	93,099	58,058	40,730
2.000 - 8 UNR	15.7	7.90	284	20,093	12,531	8,791
1.000 - 12 UNJ	13.7	6.77	243	53,082	33,087	23,208
1.000 - 8 UNJ	20.7	5.79	208	139,655	87,093	61,098
2.000 - 8 UNJ	20.7	7.34	264	31,658	19,743	13,850

NOTES: Maximum Thread Section Stress is 50% of the minimum UTS of 160 Ksi, Stress Range R = 0.10

(SD) from the mean. The 95% lower bound is equivalent to 1.645 SD. Thus, the aerospace minimum individual test life limit can be treated as a very unlikely occurrence (0.2%) based on the reference 0.625 in. (16 mm) diameter database variability.

The rolled thread size dependent fatigue life relationships, shown in Fig. 5, can be numerically expressed in the form of:

$$N \text{ (number of cycles)} = B \times \text{(basic thread diameter, BTD, in.)}^m \qquad (2)$$

where B and m are fitting constants.

The fitted constants for the minimum acceptable fatigue lifetimes of 8-UNR and 8-UNJ (3 mm pitch) thread forms of diameters larger than 1.00 in. (25 mm) are represented by (B/m) of 65 000/-2.21 and 100 000/-2.06, respectively.

Figure 5 shows the available fatigue fracture test data for each diameter thread, all of which fall to the acceptance side of the 95% lower bound curve. Thus, all large diameter threads made by controlled in-feed rolling that have been tested with fatigue lives less than 65 000, would be acceptable to these limits. Using the formulation of Eq 2, the minimum average fatigue life acceptance limits for 2.0 in. (50 mm) diameter threads would be 14 000 and 24 000 cycles, for 8-UNR and 8-UNJ (3 mm pitch), respectively.

Compared with the fatigue life of machined (cut) threads, the rolled thread studs 95% lower-bound minimum acceptance limit has a minimum fatigue life improvement of about 5× for 1.0 in. (25 mm) diameter threads, and about 2.5× for the 2.0 in. (50 mm) diameters. For a typical fatigue life, rolled thread studs are expected to have a 10× to 5× improvement, respectively, for the same thread diameters. This improvement factor over cut threads is due to the presence of residual compressive stresses in the thread roots of rolled threads. Rolled threads from small and large diameter threaded studs, with reduced fatigue acceptance lives, have exhibited significantly improved resistance over cut threads when exposed to aggressive aqueous environments prone to stress corrosion cracking [3].

Limited fatigue testing of studs having both small and large diameter thru-feed rolled threads also show equivalent or better lifetimes than in-feed threads.

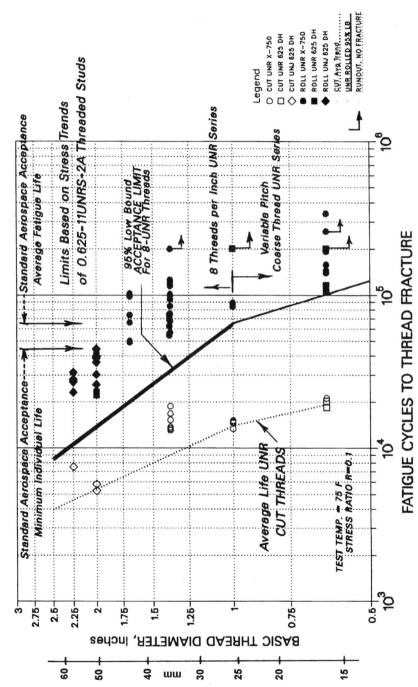

FIG. 5—Air fatigue lifetime acceptance limits for rolled threads, stressed to 50% UTS as a function of thread diameter and compared with limited cut and rolled thread test data.

Results from Fatigue Tests of Nut Geometry Variables

Undocumented commercial experience from fatigue tests employing various compression nut geometry has noted significant effects on threaded fastener fatigue test life. This led to limited tests to evaluate the principal cause of fastener fatigue life sensitivity to test nut geometry. Again, using the principle of local thread root stress concentration, two primary variables were investigated:

(a) Nut bearing surface flatness, ranging from flat to the maximum concavity allowed by the nut dimension tolerance required by MIL-STD-1312, Fig. 2 and Table II of Test 11A, and

(b) Nut thread flank angle mismatch between external and internal threads.

A simple cylindrical external nut shape was compared to the flanged type shape allowed by MIL-STD-1312, 11A, in addition to a range of nut diameters ranging from 1.3 to 2.0 times the basic thread diameter (BTD). For evaluation of these nut geometry variables, the 0.625 to 11UNR-2B (M16×2) thread form was tested using Alloy X-750 HTH for both the rolled thread test studs and the compression nuts.

Because of the smaller SCF of 5.7 for the 0.625 in. (16 mm) diameter threads and the longer fracture fatigue lifetime of about 200 000 cycles, when tested at 50% of UTS, the testing was conducted at 60% of UTS (95 ksi, 655 MPa, maximum stress) at a stress ratio of $R = 0.10$. Baseline fatigue life data for flat bearing surfaced nuts of 1.6× BTD and 2.0× BTD exhibited no difference in trends between nut diameters and averaged 66 000 cycles. The smaller diameter nut(1.3× BTD) had a shorter fracture fatigue life, averaging 40 000 cycles from two test studs. Flanged nut geometry testing, comparing a flat bearing surface to a concave surface, showed a factor of 2 to 3 longer fatigue cycle life for the concave nuts. The flat nuts had a bearing surface dimensional runout of less than 0.001 in. (0.025 mm), whereas, the concave surfaced nuts had a runout of 0.003 to 0.004 in. (0.08 to 0.10 mm), near the maximum allowed concavity of 0.004 in. (0.10 mm) per MIL-STD-1312, 11A, for the 0.625 in. (16 mm) diameter thread test nut.

Figure 6 shows the fretting surface contact footprints for bearing surfaces including the observed fatigue lives. In one test of studs and mixed geometry nuts (one flat and one concave), fracture occurred in the flat nut threads. Figure 6 shows nut-to-tensioner cup contact only at the outer nut diameter, whereas the flat surface nut had a relatively uniform contact footprint. Application on the test load at a location farther from the stud thread root with the concave nuts is believed to lower the local stress concentration in the first nut engaged screw thread, thus increasing the fatigue lifetime.

Figure 7 shows the data trends from concave and flat bearing surface flanged nuts in addition to those with mating flanks which had contact both above and below the screw thread pitchline. When the nut contact is located above the pitchline flank cracks can occur which, if sufficiently deep, can relieve the first engaged thread root stress; thereby extending the fatigue life. Both thread root cracking and fracture occurred in the set with nut flank contact above the pitchline. The average fatigue lifetime observed was similar to the set with nut flank contact below the pitchline. Thus, from this test, the most sensitive nut geometry feature affecting fatigue life is the bearing surface concavity.

While thread flank contact mismatch showed no significant effect in this test, other tests have shown significant increases in fatigue life due to contact above the pitchline with resulting large flank cracking without any thread root cracking and thread fracture. Figure 8 shows flank contact mismatch induced fatigue cracking in which fatigue runout (no thread fracture) resulted after 260 000 cycles of testing at 50% UTS for Alloy X-750 HTH. Similar testing with nuts of matched flank contact produced fatigue fracture lives ranging from

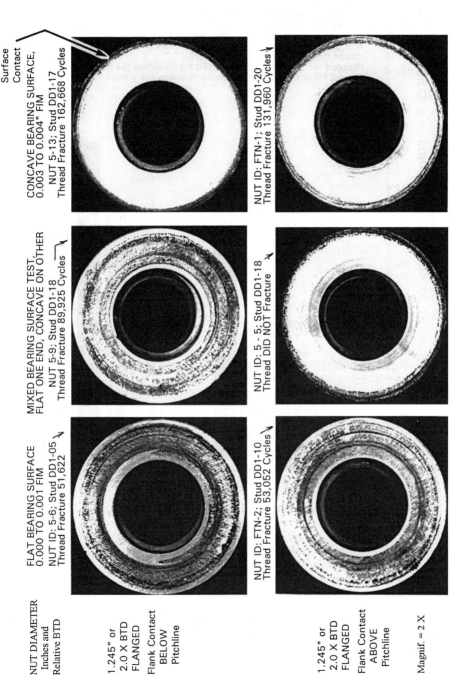

FIG. 6—*Fatigue test wear marks on bearing surface of flange design test nuts as a function of nut bearing surface flatness.*

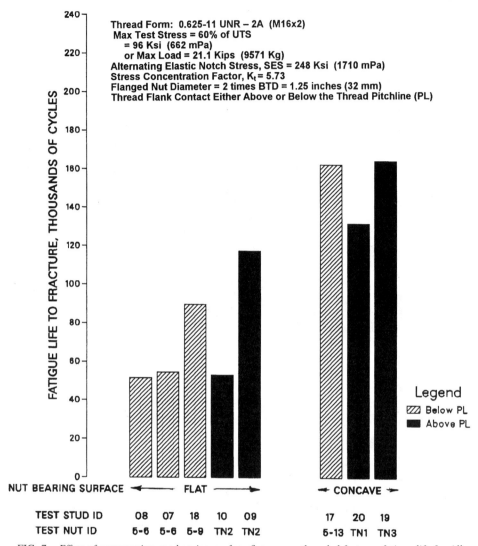

FIG. 7—*Effect of compression nut bearing surface flatness on threaded fastener fatigue life for Alloy X-750 HTH tension tested at maximum stress of 60% UTS, R = 0.1 and 75°F.*

31 000 to 115 000 cycles. To avoid inaccurate fatigue lifetime measurements, thread flank contact above the thread pitchline must be controlled.

When fatigue acceptance or process qualification tests are required by MIL-DTL-24789, the required test nut geometry is different than MIL-STD-1312, 11A. MIL-DTL-24789 requires tighter tolerances on bearing surface runout of approximately 25% of the aerospace allowances on nut concavity. Also, a special feature has been added to the internal thread to avoid external thread contact above the thread pitchline. This feature (controlled flank contact, CFC) is a metal cutout of about 0.002 in. (0.05 mm) deep at the major diameter of the internal thread and extending over the upper one-third of the thread flank. Figure 9 illustrates the nut geometry features which help avoid potential fatigue lifetime effects from nut ge-

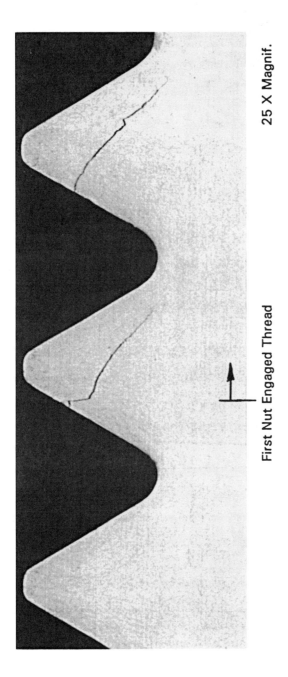

25 X Magnif.

First Nut Engaged Thread

No Root Cracks Observed After 260,000 Cycles
With Maximum Stress at 50% of UTS, 80 Ksi (552 mPa)

Thread Form: 0.625 – 11 UNR – 2A (M16x2)

FIG. 8—*Fatigue cracks in thread flanks of Alloy X-750 HTH rolled threads with engaged nut contact footprints only above the thread pitchline.*

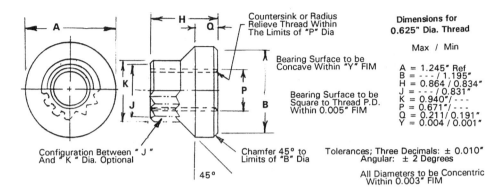

Countersink or Radius
Relieve Thread Within
The Limits of "P" Dia

Bearing Surface to be
Concave Within "Y" FIM

Bearing Surface to be
Square to Thread P.D.
Within 0.005" FIM

Configuration Between " J "
And " K " Dia. Optional

Chamfer 45° to
Limits of "B" Dia

45°

**Dimensions for
0.625" Dia. Thread**

Max / Min

A = 1.245" Ref
B = - - - / 1.195"
H = 0.864 / 0.834"
J = - - - / 0.831"
K = 0.940"/ - - -
P = 0.671"/ - - -
Q = 0.211/ 0.191"
Y = 0.004 / 0.001"

Tolerances; Three Decimals: ± 0.010"
Angular: ± 2 Degrees

All Diameters to be Concentric
Within 0.003" FIM

Flanged Fatigue Test Nut Design Allowed by MIL-STD-1312, Test 11A

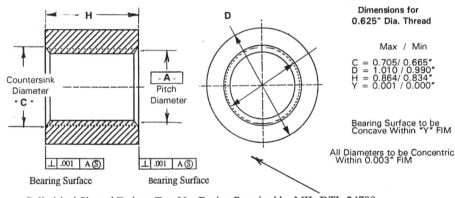

Countersink
Diameter
" C "

Pitch
Diameter

Bearing Surface   Bearing Surface

**Dimensions for
0.625" Dia. Thread**

Max / Min

C = 0.705/ 0.665"
D = 1.010 / 0.990"
H = 0.864/ 0.834"
Y = 0.001 / 0.000"

Bearing Surface to be
Concave Within "Y" FIM

All Diameters to be Concentric
Within 0.003" FIM

Cylindrical Shaped Fatigue Test Nut Design Required by MIL-DTL-24789

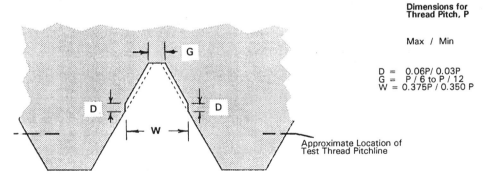

Approximate Location of
Test Thread Pitchline

**Dimensions for
Thread Pitch, P**

Max / Min

D =   0.06P/ 0.03P
G =   P / 6 to P / 12
W = 0.375P / 0.350 P

Controlled Flank Contact Feature (Cutout) of Test Nut Design of MIL-DTL-24789

FIG. 9—*Illustrations of flanged fatigue test nut on MIL-1312,Test 11A and cylindrical test nut design required by MIL-DTL-24789(SH).*

ometry. The simple cylindrical geometry and double-ended test capacity of these test nuts offset the costs of the additional features while providing more reliable test data.

Conclusions

1. Reduced fatigue lives in large-diameter rolled threads, relative to existing aerospace standards for smaller-diameter threads, can be explained by accounting for thread root notch effects, a principal factor not present in current aerospace fastener fatigue acceptance requirements.

2. Combining a fatigue database from a small-diameter stud with thread diameter and pitch-dependent elastic stress concentration factors can provide a practical formulation for thread "size" dependent fatigue acceptance limits.

3. Fatigue lifetime limits, which include thread diameter dependent stress concentration factors for 0.125 in. (3 mm) pitch threads, compare favorably with the in-feed rolled thread test database for up to 2.25 in. (60 mm) in diameter.

4. Significant effects of test nut geometry and tolerances on fatigue fracture life of threads prompted adoption of additional control features in the rolled thread procurement specification MIL-DTL-24789.

References

[1] Lin, C. S., Lourilliard, J. J., and Hood, A. C., "Stress Corrosion Cracking of High Strength Bolting," *Stress Corrosion Testing, ASTM STP 425,* 1967, pp. 84–98.
[2] Roach, T. S., "Aerospace High Performance Fasteners Resist Stress Corrosion Cracking," *Materials Performance,* Vol. 23, No. 9, Sept. 1984, p. 42.
[3] Kephart, A. R. and Hayden, S. Z., "Benefits of Thread Rolling Process to the Stress Corrosion Cracking and Fatigue Resistance of High Strength Fasteners," *Proceedings, Sixth International Symposium on Environmental Degradation of Materials in Nuclear Power Systems-Water Reactors,* 1–5 Aug. 1993, San Diego, CA.
[4] MIL-B-85604A, Bolt Nickel Alloy 718, Tension, High Strength, 125 ksi F_{sy} and 220 ksi F_{tu}, High Temperature, Spline Drive, General Specification for, Preparing Activity, Navy—AS, Project No. 5306-1204, 25 Jan. 1988.
[5] SAE-AS7466, Bolts and Screws, Nickel Alloy, Corrosion and Heat Resistant, Forged Head, Roll Threaded, Fatigue Rated, *Aerospace Standard,* Society of Automotive Engineers, Inc., FSC 5306, 22 Jan. 1991.
[6] MIL-STD-1312, Test Method 11A, Tension Fatigue, 19 Oct. 1984, FSC 53GP.
[7] MIL-DTL-24789(SH), Detail Specification, Studs; Rolled Thread, 28 Feb. 1997, FSC 5307.
[8] Crispel, C., "New Data on Fastener Fatigue," *Machine Design,* 22 April, 1982.
[9] Harvey, J. F., *Theory and Design of Modern Pressure Vessels,* Van Nostrand Reinhold Co., 1974, p. 353.

Chris B. Dawson[1] and Mark L. Thomsen[2]

Experimental Techniques to Evaluate Fatigue Crack Growth in Preflawed Bolts Under Tensile Loads

REFERENCE: Dawson, C. B. and Thomsen, M. L., **"Experimental Techniques to Evaluate Fatigue Crack Growth in Preflawed Bolts Under Tensile Loads,"** *Structural Integrity of Fasteners, Second Volume: ASTM STP 1391,* P. M. Toor, Ed., American Society for Testing and Materials, West Conshohocken, PA, 2000, pp. 162–191.

ABSTRACT: Fatigue crack growth testing of Inconel and multiphase 159 bolts is presented. The direct-current potential difference technique is shown to be successful in obtaining crack length data from the test bolts with short grip lengths. Three stress intensity solutions for round sections are presented with a preferred solution suggested. This preference is based on the ability to apply a geometric correction that is variable with crack depth and crack aspect ratio. The crack aspect ratio functionality was determined experimentally by "marker banding" the fracture surface and adjusting the stress intensity solution, following the test, by adjusting the crack aspect ratio. The fatigue crack growth of the Inconel was completely successful. The MP159 cracking was initially Mode I but quickly turned mixed mode, thus invalidating the Mode I crack growth data.

KEYWORDS: fatigue, crack growth, bolts, potential drop, stress intensity, marker band

The utilization of fasteners represents a major category of attachment techniques employed for mechanical fastening systems. A subset of fastening systems used in many critical structures such as automotive, aerospace and various other commercial transit systems are threaded fasteners. Threaded fasteners are used extensively for the attachment and critical joining of structures and components; offering the advantage of quick and easy assembly. In particular, bolts an threaded fasteners permit rapid disassembly for repairs and maintenance. However advantageous the benefits of employing threaded fasteners may be, their use must also be approached with certain precautions. A thorough understanding of the limitations within the overall structure is necessary. Analysis must be employed to ensure that the integrity of the systems and structures they support will not cause a major catastrophe in the event of fastener failures. The danger arises from crack growth within the fasteners due to repetitive loads and hostile environments. Crack growth within bolts represents a major problem because crack propagation is not normally detectable visually. When the conditions for crack propagation exist, the absence of advance-warning techniques has resulted in structural integrity loss. Therefore, accurate fracture mechanics analysis techniques of surface crack formation in bolts are needed for reliable predictions of crack growth rates and residual strengths.

[1] 14525 NE 30th Place #14-D, Bellevue, WA 98007.
[2] Lead engineer, Structural Materials Laboratory, Boeing Commercial Airplane Group, P.O. Box 3707, MC 45-12, Seattle, WA 98124-2207.

A great deal of research has been conducted in determining the nucleation of cracks in threaded fasteners due to the combination of loads and stresses at the critical thread root. Generally, cracks start at the first or last thread. The behavior of cracks nucleating at the thread roots of bolts is a relatively well-understood phenomenon. However, very little research has been conducted on fatigue crack nucleation in the shank of threaded fasteners. The main concern for fatigue cracking in the shank of bolts and threaded fasteners is due to scratches, gouges, nicks, dings, etc. associated with the installation or removal of these components during inspection and maintenance functions. Therefore, the main purpose of this research was to apply and verify a stress intensity solution for circumferential surface cracks nucleating in bolts and threaded fasteners.

The experimental approach for a stress intensity solution was developed by testing 28.6-mm (1.125 in.) and 31.8-mm-diameter (1.250 in.) bolts manufactured from Inconel 718, a nickel-based steel alloy and MP159, a cobalt-based multiphase steel alloy. To simulate surface imperfections encountered in service and installation conditions, an artificial defect, a_n, of 2 mm (0.08 in.) in depth and approximate circumferential length, $2b_n$, of 16 mm (0.64 in.) was utilized to promote cracking. This artificial defect was induced in the material by means of an electrospark discharge machine (EDM). Cracks nucleating from the EDM flaw were allowed to propagate from the initial notch length to fracture under constant-amplitude loading conditions.

To measure the growth of fatigue cracks in the threaded fastener specimens the technique of direct-current electrical potential difference method for continuous in-situ monitoring of the growth of fatigue crack propagation was employed. A constant direct current is passed through the specimens and the voltage difference is measured between two probe points, one above and the other below the midplane of the EDM flaw. As the crack grows through the specimen, the voltage difference increases due to the fact that current density about the crack is intensified. The measured voltage is converted to a cyclic depth by means of a calibration equation, which is derived empirically through experimentation. For this experiment the depth of the crack is defined along the short axis of the specimen and techniques are employed to determine the crack aspect ratio (a/b where a is crack depth and b the half-circumferential crack length) as the crack propagates through the specimen. The results were analyzed utilizing linear elastic fracture mechanics to produce average crack growth rates, nondimensional stress intensity geometric corrections, β, critical crack depths, a_{cr}, and stress intensity factors, K_I. The verification of an accurate stress intensity factor for bolts and the experimental techniques will be useful in predicting crack growth rates and fracture strengths. These data will assist in designing safe structural components and in establishing standards for inspection intervals for fasteners used in critical attachment capacities.

Experimental Details

To determine the dimensionless stress intensity factors, crack growth rates and crack front profiles, three different types of materials were tested.

1. Inconel 718 A and B, nickel-based steel alloys where ultimate strengths are 1241 MPa (180 ksi) and 1517 MPa (220 ksi) respectively
2. MP159, a cobalt-based multiphase steel alloy 1930 MPa (280 ksi)

These three types of fasteners exhibit different crack growth rates and fatigue lives. It is necessary to obtain a thorough understanding of the crack propagation characteristics of each to ensure the structural integrity of the systems they support.

Specimen Description

Nine tension bolts, three from each material type, were utilized for the experiment. All specimens tested were standard 28.6 mm (1.125 in.) and 31.8 mm (1.250 in.) tension bolts. All of the tests were performed with the specimens in the "as manufactured" state with the exception of a starter notch induced in the center of the shank. The starter notch was produced by an Eldora model D15 EDM machine with a copper electrode wire approximately 0.18 mm (0.007 in.) thick.

Direct-Current Electrical Potential Difference Method (dcEPD)

Due to the overall length of the test specimens and the fixtures needed for testing, compliance measurements were not feasible for this experiment. Therefore, the direct-current electrical potential difference (dcEPD) method of the growth of cracks within the bolts was employed. The principle of the dcEPD method is to monitor small fatigue cracks from an artificially induced crack starter notch. The voltage difference increases with the growth of the crack because current density about the crack is intensified. Measured voltages versus load cycles are converted to a cyclic crack depth by means of a calibration equation developed by Roe-Coffin [1]. This is determined either by experimentation or from analysis of the electrical field. In this experiment, a constant direct current of 5.85 A was passed through the preflawed bolts. A voltage difference of between 70 and 500 μV (depending on specimen type and crack length) was measured between the two probe points.

The crack depth is defined along the short axis of the ellipse and for a known value of the crack aspect ratio, a/b, as seen in Fig. 1. Determination of the crack aspect ratio for each specimen was ascertained by use of marker bands at specific crack depths.

Determining the depth of a crack by the dcEPD method is not an extremely complicated technique. For a uniform direct-current field, the potential V between two fixed locations that contain a crack is described by a function of current density, resistivity, crack size, shape and voltage probe location. For a particular experiment, current and metal resistivity are often not known but are constant for a particular test. For this reason, potential solutions are referenced to the results for an initial notch size a_n, where the potential is referred to as V_n and thus

$$V/V_n = f(a, a_n, L_p) \qquad (1)$$

Note that L_p, the distance between the crack centerline and the probe location, describes

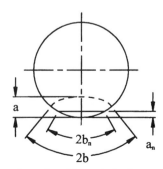

FIG. 1—*Semi-elliptical crack profile with crack depth,* a, *and circumferential length,* 2b.

the voltage probe positions. Since a notch of depth a_n is utilized to locate the site of crack propagation, the calibration must specifically account for a three-dimensional geometry. Three-dimensional cracks are those that exhibit semi-elliptical surface or corner cracks. The crack profiles in this experiment tended to be semi-elliptical.

The voltage solution for three-dimensional cracks in a uniform current field is extremely complicated. The solution must take into account that the crack has two dimensions: a depth (a) and a circumferential length, 2b. Roe and Coffin derived a potential formula for a semi-elliptical surface notch with a height 2h in an infinite body under uniform current [1]. They derived this solution from a fluid flow analysis formulated by Milne-Thompson [2] and extended and verified by Gangloff [3], Van Stone [4], and Heubaum [5]. Similar to the two-dimensional solution, it is necessary to normalize the Roe-Coffin potential solution for that of a reference crack or notch geometry and therefore

$$V/V_n = f(a, b, h, a_n, b_n, h_n, L_p) \tag{2}$$

The application of an analytical potential solution to a three-dimensional crack is vastly more complicated than a through-thickness case because the crack shape changes frequently as the crack progresses. Therefore, the equations require an iterative solution. Because of the complexity, tables of V/V_n versus crack depth must be calculated and then employed in a "look-up" appendix for computerized experiments. The Roe-Coffin potential solution for a semi-elliptical notch with a crack depth (a), a surface length 2b and a height 2h are described by

$$V = V_o L_p [\{(1 - k^2 \sin^2\theta)/\tan\theta\} + E(k, \theta) - Q]/[E(k, \pi/2) - Q] \tag{3}$$

for the case where h equals zero, and

$$V_o = \text{remote potential}$$
$$E(\delta_1, \delta_2) = \int^{\delta_2}(1 - \delta_1^2 \sin^2 \Phi)^{1/2} \, d\Phi$$
$$\theta = \tan^{-1}(\alpha)^{1/2}$$
$$\theta_o = \tan^{-1}(h/\beta)$$
$$Q = E(k, \theta_o) + h\beta^2/ab\lambda$$

There are two solutions for α, β, and λ depending upon whether the crack aspect ratio is greater or less than unity. In each case, the value of α varies depending on the relative difference between the crack dimensions and the probe spacing. The probe location is defined as $x = L_p$, $y = 0$, and z where z is the dimension by which the probe spacing is displaced from the specimen centerline:

for $b \geq a \geq h$

$$\beta = a^2 - h^2$$

$$\lambda = b^2 - h^2$$

$$k^2 = 1 - \beta^2/\lambda^2$$

if $L_p^2 + z^2 > \lambda^2$

$$\alpha = \frac{1}{2}[L_p^2 + z^2 - \lambda^2/\beta^2 + \{(L_p^2 + z^2 - \lambda^2/\beta^2)^2 + 4\lambda^2 L_p^2/\beta^4\}^{1/2}] \tag{3a}$$

$$\text{if } L_p^2 + z^2 < \lambda^2$$

$$\alpha = 2\lambda^2 L_p^2/\beta^4/[(\lambda^2 - L_p^2 - z^2/\beta^2) + \{(\lambda^2 - L_p^2 - z^2/\beta^2)^2 + 4\lambda^2 L_p^2/\beta^4\}^{1/2}] \qquad (3b)$$

$$\text{for } a \geq b \geq h$$

$$\beta = b^2 - h^2$$

$$\lambda = a^2 - h^2$$

$$k^2 = 1 - \beta^2/\lambda^2$$

$$\text{if } L_p^2 + z^2 > \beta^2$$

$$\alpha = \tfrac{1}{2}[(L_p^2 + z^2)/\beta^2 - 1 + \{(L_p^2 + z^2/\beta^2 - 1)^2 + 4L_p^2/\beta^2\}^{1/2}] \qquad (3c)$$

$$\text{if } L_p^2 + z^2 < \beta^2$$

$$\alpha = 2L_p^2/\beta^2/[(1 - L_p^2 + z^2/\beta^2) + \{(1 - L_p^2 + z^2/\beta^2)^2 + 4L_p^2/\beta^2\}^{1/2}] \qquad (3d)$$

The effect of the starter notch on the potential field is addressed by applying the following algorithm from Gangloff [3]

$$V = V_{ba} + Q_o(V_n - V_{bn}) \qquad (4)$$

where

V_n = potential for notch with dimensions $a_n \times 2b_n \times 2h_n$
V_{ba} = potential for crack with dimensions $a \times 2b(a \geq a_n, b \geq b_n, h = 0)$
V_{bn} = potential for crack with notch dimensions $a_n \times 2b_n(h = 0)$

Q_o is an empirical function that guarantees $V/V_n = 0$ for $a = a_n$ and it decays linearly to zero as the crack depth reaches twice the notch depth. Therefore

$$Q_o = (2 - a/a_n) \text{ for } a_n \leq a \leq 2a_n \qquad (4a)$$

$$Q_o = 0 \text{ for } a \geq 2a_n \qquad (4b)$$

By predicting the relationship between crack size and voltage with the equations referenced above, it is possible to assess the applied direct-current and voltage resolution needed for any small crack geometry, specimen size and alloy. A single voltage measured for a specific crack and probe geometry is used with model computations to define absolute voltages for any crack size and probe location. To establish the effects of material on the potential as a function of crack size, potential values and current density are linearly scaled with known electrical resistivity values and by the applied current for specific specimen cross-sectional areas.

dcEPD Equipment Description

Instrumentation utilized for the dc potential difference method included a Tektronix model PD501 Fracture Systems Research Current Amplifier, a Tektronix model PD502 Fracture Systems Research Dual PD Amplifier power supply and a computer analog-to-digital board with controlling software developed by Fracture Technology Associates. The complete hardware and software integrated system is set up to interface with a closed-loop, servo-hydraulic load frame. It is configured for fatigue crack growth rate testing with instrumentation to measure electrical potential differences. The system required four input signals to the test system. These include the load signal, direct-current electrical potential difference signal, command signal provided by the computer system and a digital or analog control system utilized to turn the current on, off and to reverse it. The command and feedback signals assume full scale of ± 10 V. A digital I/O is utilized to monitor the status of the test machine hydraulics. The PD502 20 V, 20 A dc power supply along with two PD501 differential amplifiers with selectable filters and gains from 500 to 50 000 are used to supply the required current and resulting output voltage, respectively.

Figure 2 shows a schematic diagram of the dcEPD circuit coupled with the computer-based data acquisition system and the servo-hydraulic test machine. A power relay switch is used to switch the polarity of the applied current to eliminate thermally induced voltages. The current switching interval was determined experimentally and functioned most effectively at an interval of 0.1 s. The switching interval is automatically controlled by the digital-

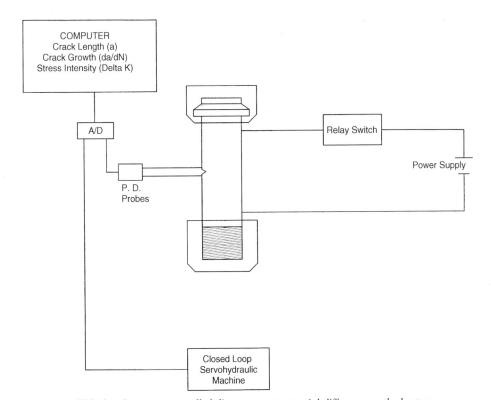

FIG. 2—*Computer-controlled direct-current potential difference method setup.*

to-analog signal dictated from the computer software. The gain amplifier was set for a gain of 5000 for the potential difference readings. These voltage readings were interpreted by the software and compared with the look-up appendix developed from Eqs 2 to 4 to determine the appropriate crack depth.

Current was applied to the specimen far from the crack with aluminum current leads that were attached to the specimens by means of spot welding. The current leads are applied directly to the specimen sufficiently far from the surface flaw to provide a uniform current density and so as not to provide sites for fatigue cracking. Again, critical to the dcEPD measurement is the attachment of the voltage probes to measure the small increase in potential associated with crack extension. Accurate placement of the leads relative to the EDM flaw that will produce reliable electrical conductivity and structural integrity was achieved by spot welding. The attachment method was chosen such that it did not introduce metallurgical heterogeneity or stress concentrations that could result in fatigue cracks. Figure 3 shows the probe placements for measuring the changes in voltages as the crack propagates from the original EDM flaw through the shank of the specimens.

The use of dcEPD monitoring of the growth of three-dimensional cracks is quite complicated if the crack aspect ratio varies during crack extension. The solutions for electrical potential difference measurements for three-dimensional cracks presented in Eqs 2 and 3 describe an infinite number of crack sizes and shapes at a given potential. It is, however, necessary to know the crack aspect ratio in order to uniquely determine the crack depth. This was accomplished by marker banding the test specimens at various depths to ensure the accuracy of the system. Several marker band cycles were performed on each specimen to determine the crack aspect ratio at various depths. By monitoring the changes in crack shape in this way, the data were used to develop a crack shape relationship and to analyze the results of the tests.

Test Fixtures and Test Procedures

Since the specimens could not be gripped directly into the test frame, it was necessary to develop fixtures that would accommodate both the test specimens and the direct-current potential difference measurement probes. A "split ring" fixture was used to test the bolt specimens since this design had proven to be successful for fatigue testing of threaded

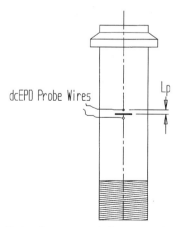

FIG. 3—*Probe placements relative to initial EDM flaw.*

fasteners for quality assurance purposes. This design was chosen because only tensile loads were applied and the design minimized the amount of bending in the specimens. Bending of any type, either by misalignment of the test frame or bending moments due to fixturing, were considered undesirable. This split-ring design for fatigue tests of bolts reduced bending stresses to a maximum value of 3% of the tensile stress. This was determined by strain gage alignment procedures per ASTM Standard Practice for Verification of Specimen Alignment Under Tensile Loading (E 1012).

Test Details

All testing was performed on a 500 kN (100 kips) servo-hydraulic load frame at room temperatures and ambient humidity. Figures 4 and 5 show the test setup and a close-up of the probes for the direct-current electrical potential measurement method employed in the test. The load frame was calibrated per ASTM Standard Practice for Force Verification of Testing Machines (E 4) using equipment traceable to the National Bureau of Standards. The specimens were tested under constant-amplitude cyclic tension loads at a frequency of 10 Hz. The test parameter levels can be seen in Table 1. Testing of all threaded fasteners adhered to ASTM Standard Test Method for Measurement of Fatigue Crack Growth Rates (E 647), where applicable.

1.25 inch

FIG. 4—*Potential drop probe placement.*

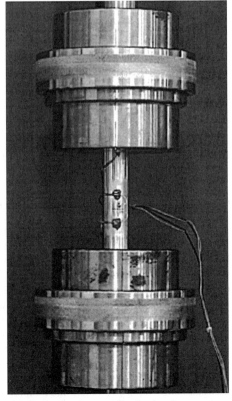

FIG. 5—*Specimen and split-ring fixture setup.*

TABLE 1—*Test parameter matrix.*

Material	Spec. ID	Diameter, mm (in.)	Area, mm² (in.²)	Max. Stress, MPa (ksi)	Max. Test Load, kg (kips)	Min. Test Load, kg (kips)
Inconel 718-A	PN1	31.750 (1.250)	791.25 (1.227)	216.5 (31.4)	17 477 (38.5)	1748 (3.85)
	PN2	31.750 (1.250)	791.75 (1.227)	307.2 (44.6)	24 800 (54.7)	2480 (5.47)
	PN3	31.750 (1.250)	791.75 (1.227)	386.1 (56.0)	30 721 (68.7)	3072 (6.87)
Inconel 718-B	US1	28.575 (1.125)	641.30 (0.994)	216.5 (31.4)	14 156 (31.2)	1416 (3.12)
	US2	28.575 (1.125)	641.30 (0.994)	307.2 (44.6)	20 087 (44.3)	2009 (4.43)
	US3	28.575 (1.125)	641.30 (0.994)	386.1 (56.0)	25 252 (55.7)	2525 (5.57)
MP159	TB03	28.575 (1.125)	641.30 (0.994)	216.5 (31.4)	14 156 (31.2)	1416 (3.12)
	TB11	28.575 (1.125)	641.30 (0.994)	216.5 (31.4)	14 156 (31.2)	1416 (3.12)
	TB17	31.750 (1.250)	791.72 (1.227)	307.2 (44.6)	24 800 (54.7)	2480 (5.47)

The Roe-Coffin equations were used, with a probe spacing L_p = 3.8 mm (0.15 in.), to develop a look-up appendix that could provide continuous monitoring of the crack growth. The specimens were initially loaded with small tensile loads that would not result in crack growth from the notch. This load was determined through trial and error and was concluded to work best at 4536 kg (10 000 lbs). This was used as a reference load to calibrate the system so that the look-up appendix could provide accurate crack depth readings. Ten electrical voltage potential readings were taken and averaged to provide the potential drop reference reading at the EDM flaw depth of 2 mm (0.080 in.). After obtaining the initial potential voltage reading at the EDM flaw the bolts were allowed to cycle at the test stress levels.

The technique of marker banding was performed on all specimens at different increments of crack growth to obtain more data on the crack profile of each bolt tested. After fracture, the fatigue crack surfaces were photographed. The marker band technique is simply the application of marker cycles at selected intervals throughout the test to characterize the fracture surface. Upon completion of the test, visual measurements were made of the marker bands on the fracture surface. These bands were then correlated to a corresponding number of cycles to obtain crack growth information and, more important, they provided the shape of the crack as it progressed.

Generally, marking the fracture surface can be accomplished by changing the stress ratio for a select number of cycles at intervals throughout the tests. The stress ratio was changed from R = 0.10 to R = 0.70 (holding the maximum load constant) and a reliable mark of the fracture surface was created without affecting the overall material behavior in the tests. Two other parameters, which are important in obtaining a successful marker band on the fracture surface, are the number of marker cycles and the number of test cycles applied between bands. This was determined experimentally for each of the material types. For each of the specimens in the test, two marker bands were applied. Each marker band was applied for a maximum of 2000 cycles at depths of approximately 6 mm (0.24 in.) and 8 mm (0.34 in.). The first marker band depth was applied because at this value the entire EDM flaw was engulfed by the crack. The second marker band depth was chosen so that it had no effect on the crack growth rate as the flaw progressed closer to the critical depth. This provided four crack aspect ratios for each specimen. The initial elliptical aspect ratio was zero for a straight-edge crack, then two additional aspect ratios were produced by marker band methods and one was determined at the fracture depth. The crack aspect ratio at fracture was utilized for the polynomial required by the software and the others were used when interpreting the results after the tests concluded.

Data Acquisition and Analysis

Load, crack depth, maximum K, ΔK, crack growth rates, electrical potential readings and cycle count for all tests were recorded by using an AGI model EX-3000D computer programmed specifically for this series of tests. The software calculates the crack depths by referring to the look-up appendix, which computes the values at specific potential voltage readings. The computer software requires a polynomial solution be input for the stress intensity factor geometric correction, β. The equation developed by Shih and Chen [6] for cracks originating from the center point of the cracked shaft was employed for calculating β. The data processor uses the values at the specific crack depths as determined from the look-up appendix to estimate values for the stress intensity and ΔK. To use the Shih/Chen equation with the computer program, it was necessary to know the crack aspect ratio at fracture since this provided the most accurate readings as determined by trial and error. The elliptical crack aspect ratio initially was set at zero because the EDM flaw is a straight-edge crack. As the crack progresses through the diameter it becomes semi-elliptical and eventually converges to a constant value similar to an equation presented by Carpinteri and Brighenti [7]. The software is able to compute only values for the crack depth, a, and cannot track the variation in the crack depth to length ratio. Therefore, it was necessary to determine the crack aspect ratios for each of the three materials. This was accomplished using the marker band technique described earlier.

Results

A total of nine threaded fasteners manufactured from three separate materials were tested at various stress levels to determine the fatigue life and crack profiles of each. The results of these tests for each case are presented in the following sections in terms of the stress intensity factors, dimensionless stress intensity factors and crack growth rates. Fractographs of the fracture surfaces are presented to show the fracture morphology.

Crack Aspect Ratios

A typical Inconel material fatigue fractured test specimen is shown in Fig. 6. The fracture origin for this Inconel specimen is normal to the bolt axis. Marker bands were applied after the cracks had exceeded the edges of the initial EDM flaw and then at various increments. Typically, two marker bands were applied per specimen. Generally these were applied early in crack advancement, so there was no influence on subsequent crack growth rates. The measured crack depths, crack lengths and aspect ratios are given in Table 2.

The crack aspect ratios at failure for the Inconel specimens were compared with those in the theory presented by Carpinteri and Brighenti [7]. Their theory concluded that the solution for the crack aspect ratio, α, tended to converge to an inclined asymptote defined by

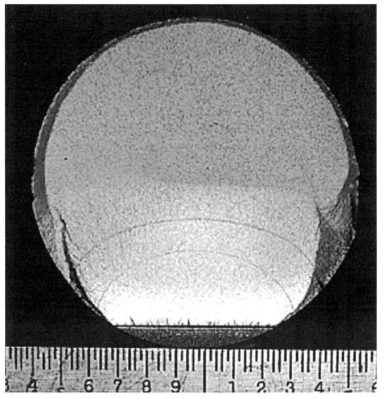

FIG. 6—*Typical Inconel fatigue test specimen fracture surface.*

$$\alpha = -1.144\xi + 1.365 \tag{5}$$

Note α is equal to approximately 0.60 to 0.70 when the fitting parameter $\xi = 0.60$. For this research, this formula did not prove to be accurate when comparing it to the experimental results. Their main premise was that the aspect ratio falls within the range of 0.60–0.70 when $a/D = 0.60$. For all Inconel specimens, the crack depth to diameter ratio did not equal 0.60. Therefore, the formula is not relevant for the results obtained in this experiment. However, the Inconel specimens did tend to converge to an aspect ratio average of 0.77 at failure. The marker bands applied after "absorbing," or extending beyond the extent of, the EDM flaw show that the aspect ratios tend to converge to these values early in the crack growth process.

Marker banding was attempted on the MP159 specimens with unsuccessful results. Because of the crack propagation characteristics, marker banding could not be achieved at any point in the growth process. Also, because of the steep angle of the fracture face and mixed-mode crack attributes, there was no crack aspect ratio associated with failure. For specimens TB03 and TB11, there was a Mode I crack extension zone normal to the bolt axis before exhibiting mixed mode loading. This zone was approximately 11.43 mm (0.45 in.) and 10.67 mm (0.42 in.) respectively.

From the marker band techniques employed for determining the crack shape profile, some general assumptions can now be presented. At the early stages of crack growth, the shape

TABLE 2—*Fatigue test results.*

Material	Spec. ID	1st Depth, mm (in.)	2nd Depth, mm (in.)	Critical Depth, a_{cr} mm (in.)	1st Band, $2b$, mm (in.)	2nd Band, $2b$, mm (in.)	Length Failure, $2b$, mm (in.)	Aspect Ratio, 1st Band	Aspect Ratio, 2nd Band	Aspect Ratio, Failure	Cycles to Failure
Inconel 718-A	PN1	6.35 (0.25)	8.64 (0.34)	20.83 (0.82)	17.78 (0.70)	22.61 (0.89)	53.34 (2.10)	0.714	0.764	0.781	227 152
	PN2	...	...	18.03 (0.71)	...	...	46.74 (1.84)	...	...	0.772	59 475
	PN3	6.35 (0.25)	9.14 (0.36)	16.00 (0.63)	18.03 (0.71)	24.38 (0.96)	42.16 (1.66)	0.704	0.750	0.759	34 732
Inconel 718-B	US1	8.38 (0.33)	10.92 (0.43)	14.22 (0.56)	22.86 (0.90)	29.21 (1.15)	27.94 (1.46)	0.733	0.748	0.767	314 585
	US2	6.35 (0.25)	9.40	11.18 (0.44)	17.27 (0.68)	25.15 (0.99)	29.21 (1.15)	0.735	0.747	0.765	116 716
	US3	6.35 (0.25)	(0.37)	8.89 (0.35)	17.53 (0.69)	...	23.37 (0.92)	0.725	...	0.761	45 905
	TB03	11.43 (0.45)	...	...	24.05 (0.947)	...	...	0.475	...	...	807 211
MP159	TB11	10.668 (0.42)	...	...	22.61 (0.89)	...	...	0.472	...	...	718 531
	TB17	...	...	...	...	...	...	...	...	...	239 626

development is strongly dependent on the initial crack configuration. As the crack develops, the shape of the crack fronts for the two different Inconel materials are similar. Also, examining the fracture surfaces of the Inconel specimens it can be seen that as the crack propagates the crack front gradually becomes flat. At the root of the EDM flaw, multiple cracks develop and join to form a smooth semi-elliptical crack front relatively quickly. This can be seen in Fig. 6. Once the crack front becomes semi-elliptical, the cracks show a tendency to grow toward a preferred profile.

Dimensionless Stress Intensity Factor Geometric Correction

The stress intensity solution for a round section is presented by Shih and Chen [6] in the standard form given by

$$K_1 = \sigma\sqrt{\pi a} \cdot \beta \qquad (6)$$

where

σ = far-field stress,
a = crack depth, and
β = geometric correction factor.

The formula they developed includes the crack depth to diameter ratio, α, and the crack aspect ratio, a/b. It accounts for both of these ratios as well as for the location of the crack origins. For cracks originating at the center of the EDM flaw for a round section in uniform tension, the correction factor is given as

$$\beta = 0.67 - 0.033(a/b) + 5.73\alpha - 0.29(a/b)^2 - 2.943(a/b)\alpha - 22.692\alpha^2$$

$$+ 2.41(a/b)^2\alpha + 10.684(a/b)\alpha^2 + 49.34\alpha^3 - 8.82(a/b)^2\alpha^2$$

$$-10.16(a/b)\alpha^3 - 21.43\alpha^4 \quad (7)$$

The geometric correction factor increases significantly with increasing crack depth ratio and decreases slightly with increasing crack aspect ratio. These corrections, with different crack aspect ratios under axial loading, are shown in Fig. 7. Figure 8 represents these corrections with different crack depth ratios. These figures are valid only for the Inconel specimens since crack aspect ratios and crack depth-to-diameter ratios could not be obtained for the MP159 specimens.

The assumption that the cracks developed from the center and not from the end points of the EDM flaw is verified by the experimental results. This formula was used since the crack aspect ratio changes with different crack depth ratio in the fatigue crack growth process.

Crack Extension Curves

Knowledge of the extension of fatigue cracks is necessary so that inspection intervals can be developed for each material type. Critical crack depths result in fracture in service and thus it is imperative to determine permissible crack sizes. Therefore, one of the main objectives of damage tolerance analysis is to develop crack extension curves depicting growth under the action of normal service fatigue loading. Figures 9 through 11 show the fatigue crack extension curves under constant-amplitude loading and the same initial crack depth.

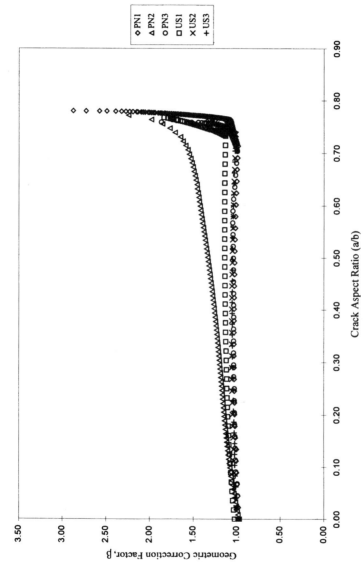

FIG. 7—*Stress intensity factor geometric correction, β, versus crack aspect ratio.*

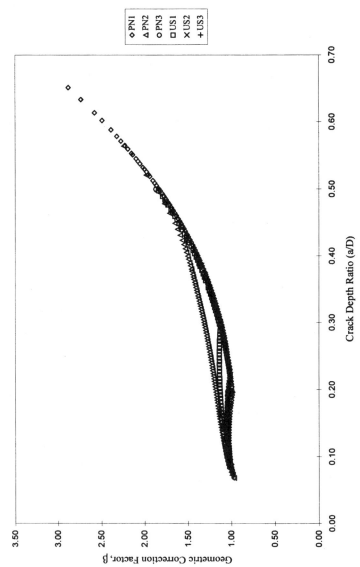

FIG. 8—Stress intensity factor geometric correction, β, versus crack depth ratio.

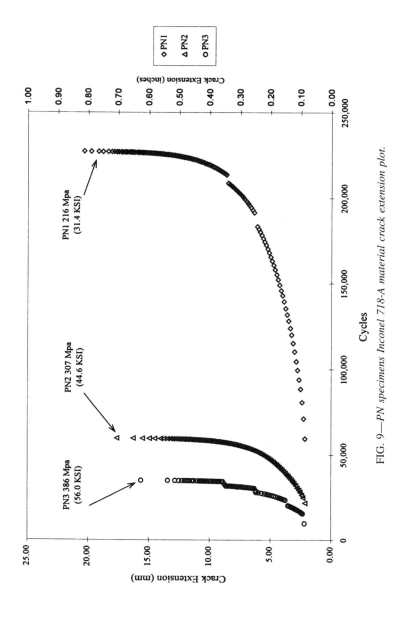

FIG. 9—*PN specimens Inconel 718-A material crack extension plot.*

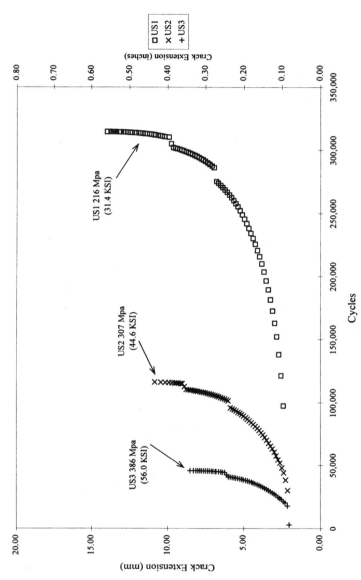

FIG. 10—US specimens Inconel 718-B material crack extension plot.

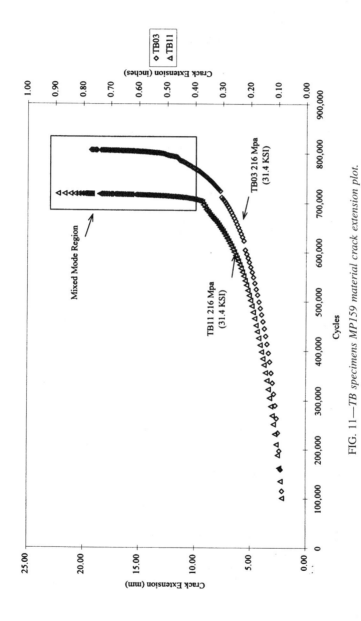

FIG. 11—*TB specimens MP159 material crack extension plot.*

Crack extension curves were constructed for the Inconel materials from the initial EDM flaw to failure. The crack extension curves for MP159 material were constructed over the entire range of data obtained. However, the specimens did not adhere to the crack planarity requirements of ASTM Standard Test Method for Measurement of Fatigue Crack Growth Rates (E 647). This condition is required for Mode I fatigue testing; therefore only part of the crack extension range is valid. The range from the initial flaw depth of 2 mm (0.08 in.) to approximately 11 mm (0.420 in.) is considered to be accurate. At the end of the Mode I region the data were accurate to approximately 1%. All values obtained after this zone are beyond the interpretation of the potential-difference software.

Fatigue Crack Growth Rates

The total cycles to failure of all test specimens are given in Table 2. Basic material fatigue crack growth data were obtained by testing three specimens from each Inconel material group. The research has shown, experimentally, that the front of a surface flaw in a metallic round bar or threaded fastener can be modeled accurately by assuming a semi-elliptical arc throughout the entire fatigue growth. The research has also proven that the crack aspect ratio changes during cyclic loading and this has a marked influence on the fatigue crack propagation characteristics. The data can then be examined by means of a two-parameter model based on the work of Paris and Erdogan. The ratio da/dN is defined as the crack propagation rate, ΔK_I is the stress intensity factor range, and A and m are the data fitting parameters. Typical crack growth rates versus ΔK_I values are plotted for the two Inconel materials in Figs. 12 and 13. The effect of the marker banding technique can be seen in these figures as the few data points that are not part of the typical sigmoidal-shaped curves.

The results of the data were analyzed to obtain an appropriate fatigue crack growth equation for each material class. A two-parameter model based on the Paris-Erdogan data fit resulted in the following growth rates

$$da/dN = 1.73 \times 10^{-13}(\Delta K_I)^{3.85}$$

$$(da/dN = 9.80 \times 10^{-12}(\Delta K_I)^{3.85}) \tag{8}$$

for Inconel 718-A material. Inconel 718-B material had a calculated fatigue growth rate defined as:

$$da/dN = 3.75 \times 10^{-15}(\Delta K_I)^{4.79}$$

$$(da/dN = 2.32 \times 10^{-13}(\Delta K_I)^{4.79}) \tag{9}$$

Crack growth data fitting for the MP159 bolts was not attempted due to the irregular crack propagation attributes of the material. These characteristics preclude the development of accurate crack growth data for the material using these particular specimens. However, the MP159 specimens did exhibit a much greater total fatigue life than the Inconel bolts.

Fatigue and Fracture Surfaces

Photographs of the fracture surface can be seen in Figs. 14 through 16. The MP159 bolts exhibited a fracture face approximately 60 deg out of the plane normal to the bolt axis. Figures 17 and 18 exhibit the steep angle associated with crack growth in the MP159 spec-

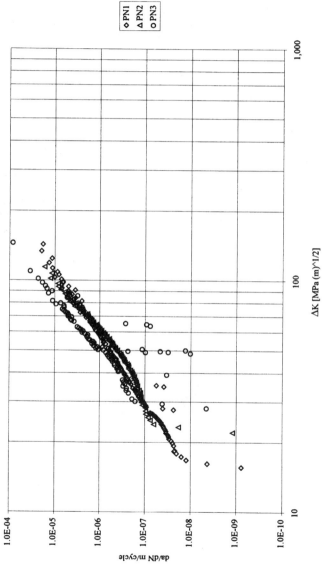

FIG. 12—*PN specimens Inconel 718-A material crack propagation rate curve.*

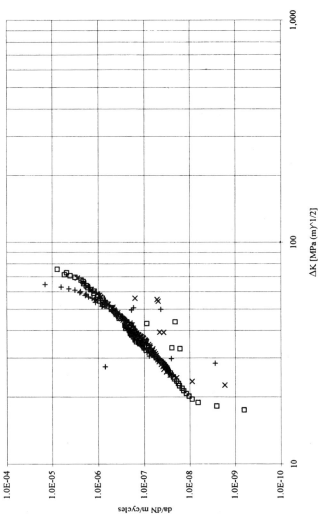

FIG. 13—US specimens Inconel 718-B material crack propagation rate curve.

FIG. 14—*Typical Inconel A fracture surface.*

imens. Several theories are considered that possibly influence this crack angularity feature. These theories include the following attributes:

1. A highly textured microstructure associated with the hexagonal closed-packed phase, which forms in thin platelets. These platelets vary in size from 20 to 1000 Å in thickness on the (111) planes of the face-centered cubic matrix.
2. Work strengthening by mechanical deformation to induce the high strength properties.
3. ΔK-dependent property, which at higher increments of ΔK promotes the steep crack growth feature.

Further tests are required to accurately determine the nature of the crack extension characteristics of multiphase materials. Other researchers have encountered similar findings in investigating multiphase cobalt materials. The research conducted by Henkener and Forman [8] encountered similar results in testing MP35 material that is similar in composition to the material investigated in the present experiment.

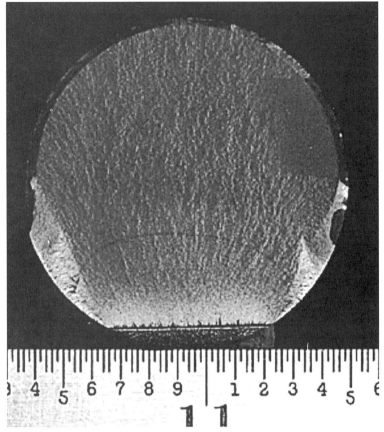

FIG. 15—*Typical Inconel B fracture surface.*

Discussion

Direct-Current Potential Difference Method for Monitoring Fatigue Cracks

The dcEPD method is a valid means to quantitatively and continuously monitor the growth kinetics of fatigue cracks in cylindrical specimens such as bolts. However, the crack propagation site must be predetermined by means of an artificially induced defect and is only valid for single Mode I cracks. The Roe-Coffin [1] equation provides a closed-form analytical model that accurately equates crack size to measured voltage for semi-elliptical surface cracks in cylindrical specimens. The method provides in-situ monitoring of the growth of small cracks and adheres to ASTM Method E 647, where applicable. Also, operator and maintenance time are minimal, instrumentation is relatively inexpensive, and operator judgment is of secondary importance.

However, the method does have some limitations that influence the effectiveness of the procedure. These include the requirement to predetermine and locally probe the crack nucleation site, the effectiveness of calibrating the system at the initial flaw, and the need to define the crack shape by independent means. A major limiting factor for the dcEPD method is the fact that it cannot detect or monitor naturally nucleated fatigue cracks where the crack

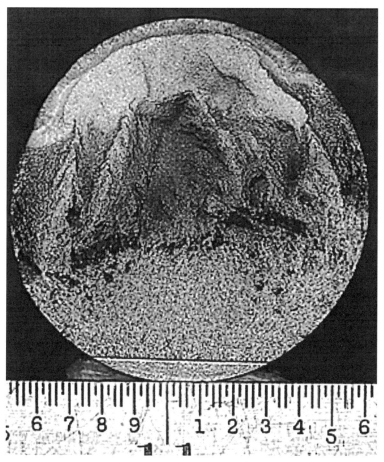

FIG. 16—*Typical MP159 fracture surface.*

location is uncertain or where multiple crack nucleation sites are present. The current research encountered multiple crack nucleation sites at the EDM flaw and initially this confounded the analytical calibration. Fortunately, the multiple cracks linked rather rapidly for this experiment and this did not have a major effect on the accuracy of the readings.

Dimensionless Stress Intensity Factor Geometric Correction

Utilization of the marker banding technique to determine the crack aspect ratio and crack front profile proved to be a valid procedure that did not effect the growth rate of the materials. This method was highly successful on both Inconel materials. The crack planarity of the MP159 specimens was not in the plane normal to the bolt's axis and, thus, marker bands could not be applied. Additionally, the morphology of the fracture surface was not supportive of this technique.

For the Inconel specimens, the crack aspect ratios and profiles could be examined and compared with the solutions proposed by other researchers. Carpinteri and Brighenti [7,9] proposed that the aspect ratio converges to an inclined asymptote. Although the solution

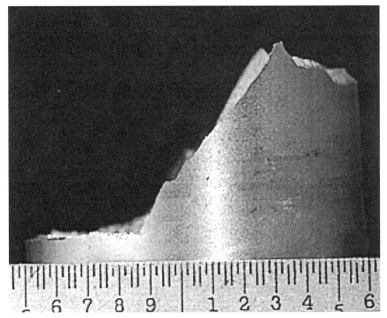

FIG. 17—*MP159 fracture surface side view.*

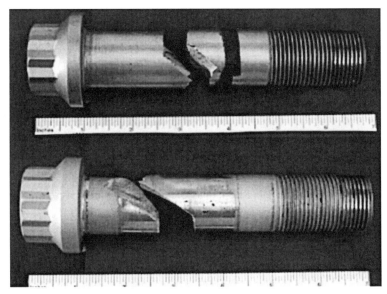

FIG. 18—*MP159 material failure comparison.*

they presented does not match the experimental research presented here, the aspect ratio did tend to converge to a stable value relatively fast in the growth process. A solution presented by Shih and Chen [6] was used in the determination of the geometric correction factor. Because of the effect of both a/b and a/D (α), this correction factor is constantly changing, thus having an effect on the stress intensity solution. By marker banding, the aspect ratio functionality was determined at various depths and used to calculate the stress intensity factor.

Many researchers present formulas that effectively ignore the aspect ratio and take into account only the crack depth ratio. Thus, it is suggested from the results of this experiment that the simulation of straight-edge or single-parameter semi-elliptical cracks for the geometric factors influencing the crack growth is not reasonable. Therefore, Shih and Chen's simulation of a two-parameter semi-elliptical arc is accurate. As here, the crack aspect ratio is continuously changing in the fatigue crack growth process and the two-parameter formula is needed for accommodating fatigue crack growth analysis. Examples of other solutions [10,11] are compared in Fig. 19 with the Shih and Chen solution. The aspect ratio shown is that associated with 0.77 found in this work. Reducing this aspect ratio, a/b, effectively increases β so it approaches the solution of James for high a/D (α) values.

Fatigue Life Comparison of Test Specimens

Each material type of the tested specimens is used in attachment applications for large-scale structural components. The MP159 specimen lives proved to be superior to those of either of the Inconel materials. In comparing the materials, the MP159 material had a 3.2 times greater life than that of the Inconel 718-A specimen and 2.3 times that of a comparable Inconel 718-B specimen. In addition, comparing the Inconel specimens, the higher-strength material possessed an average life of 1.6 times greater than that of the lower-strength specimens. At all stress levels, the Inconel 718-B specimens outperformed the Inconel 718-A specimens. In general for this research, the higher-strength materials possessed a greater total fatigue life than those of lower strengths.

Comparatively, the Inconel materials exhibited differences in the calculated fatigue crack growth rates. However, analysis of the data appears to support some similarities in the fatigue crack growth rate in Region I of the crack growth curves. Inconel 718-A material had an average minimum ΔK value equal to 22.0 MPa $(m)^{1/2}$ (20 ksi $(in.)^{1/2}$) and Inconel 718-B had an average minimum ΔK value of 22.5 MPa $(m)^{1/2}$ (20.5 ksi $(in.)^{1/2}$). For small ΔK, the crack growth is largely influenced by the microstructure of the material, the mean stress and the environment. Since the materials are similar in composition and were tested under identical conditions, this behavior is consistent. However, in Regions II and III the Inconel materials exhibited different characteristics. Inconel 718-A had an average ΔK of 145.0 MPa $(m)^{1/2}$ (132.0 ksi $(in.)^{1/2}$) and Inconel 718-B an average of 70.0 MPa $(m)^{1/2}$ (63.5 ksi $(in.)^{1/2}$) at fracture. In this region the stress intensity factor is highly affected by the material microstructure, mean stress and the thickness of the specimen. Since the test conditions and environments are identical, it can be surmised that the microstructural difference between the materials has a significant effect on the growth rates as the crack extends.

It should be noted that the environmental conditions during testing were laboratory air, i.e., non-hostile compared with the service environment. Typically, the bolt types tested in this research would be exposed to a variety of temperature variations and chemical environments, resulting in damage not accounted for in this study.

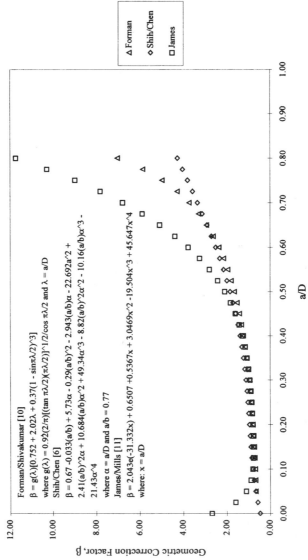

FIG. 19—Comparisons of different stress intensity geometric corrections.

Fatigue Failure Mechanisms

The fatigue process originates with the accumulation of damage at localized regions due to alternating loads. This eventually leads to the formation of cracks and their subsequent propagation. When one of the cracks has grown to such an extent that the remaining net section is insufficient to carry the applied loads, sudden fracture takes place. Thus, when observing a fatigue-fractured surface, three distinguishing features are present:

- nucleation sites
- a crack growth surface area with distinct features
- a final fractured surface

The fractographs of the test specimens in Figs. 6, 14 and 15 show that multiple crack sites originate at the artificially induced flaws. For both Inconel materials, the specimens exhibit multiple crack nucleation sites approximately perpendicular to the EDM flaw. At some point in the propagation process the multiple flaws "unite" and extend inward with a semi-elliptical crack front. It was not possible in this research to determine the depth where the multiple cracks joined to exhibit the profiles apparent at the marker bands. However, it is thought to be relatively early in the process.

For the MP159 material, it was not possible to determine if multiple cracks existed at the EDM flaw. The structure of the material is such that it is difficult to determine the origin of the cracks. While it was not possible to determine the crack origins, specifically, a few observations can be suggested. Test specimens TB03 and TB11 do display a moderately semi-elliptical crack front in the Mode I region before the crack turns out of plane. These specimens were tested at the same stress level and their characteristics are similar. In addition, during the test of each of these specimens the test was periodically halted so the specimens could be observed. Cracks developed approximately in the center of the EDM flaw for both specimens.

In comparison, the fractographs display distinct features for each material. For the Inconel specimens, a region is visible that shows multiple cracks nucleating perpendicular to the EDM flaw. This is followed by a "smooth" region that shows the progression of the crack through the diameter of the specimen. This region is created by the cyclic opening and closing of the crack and is characterized by having a level, polished appearance. The marker bands are visible in this region and, at the circumference of the specimens, shear lips can be seen on either side of the specimens. As the crack extends further into the diameter, the material exhibits a rougher texture. At fracture, the specimens exhibit a cup and cone fracture since the remaining net section is not sufficient to carry the loads. There is a visible band similar to the applied marker bands that separates the fatigue crack growth section from the sudden fracture section. This is visible in the fractographs of Figs. 6, 14 and 15. The remaining surface at fracture features a rough, dimpled appearance consistent with ductile tearing typical at rupture. The MP159 material did not display the same characteristics as the Inconel materials. The highly textured grain structure of the multiphase materials make it difficult to determine any distinct features associated with the fracture surface of the specimens. In comparison with the out-of-plane crack features of the MP159 material, the in-plane Mode I region indicates a more level and smooth surface than that of the mixed-mode sections.

Conclusions

The significance of this research was to focus attention on the material properties of the fasteners used in service fatigue loading conditions. Under the loads stated in the study, the basic material properties of the test specimens were investigated and analyzed to aid in the determination of service lives for the bolts used in critical attachment applications. Based on this experimental investigation the following conclusions are drawn:

1. The test procedures, fixtures, and techniques are valid for testing threaded fasteners and adhered to ASTM Method E 647 where applicable. The test techniques were successfully applied in testing specimens that could not be directly gripped in the load frame. The tests characterized structural components and allowed for fatigue crack growth rates to be developed.

2. The use of the direct current-electrical potential difference (dcEPD) method is a proven means to continuously monitor the growth of Mode I fatigue cracks. The technique is applicable to a wide variety of materials and geometries and provides in-situ monitoring of the growth of small cracks. A closed-form analytical model related crack size to measured voltage for surface fatigue straight-edge and semi-elliptical cracks. It is necessary to accurately calibrate the system to the initial flaw to obtain precise readings throughout the test. It is also essential to employ a preexisting machined defect to locate crack nucleation locations for the placement of the probes. The crack shape must be determined from experimentation to provide valid results. Although the technique was not successful in monitoring cracks with mixed-mode characteristics, procedures could possibly be developed to remedy this situation, i.e., utilization of multiple probes and reference probes.

3. The threaded fasteners manufactured from MP159 material possess a much greater total fatigue life than those manufactured from the Inconel materials. Generally, the specimens with higher strengths outperformed those of lower strengths for all categories.

4. The stress intensity factor geometric correction in circular specimens must be determined by accounting for the crack-to-diameter ratio and the crack aspect ratio. The formulas presented by Shih and Chen [6] accurately depict these. The marker band technique shows that the crack profile is semi-elliptical and tends to converge to a constant crack aspect ratio rapidly in the extension of the cracks.

Recommendations

1. Perform fracture toughness precracking and crack growth rate testing under a negative K gradient control or constant K control in order to maintain crack planarity.

2. Define the initial notch shape as a semi-ellipse with an aspect ratio of 0.77. This will minimize the added difficulty of determining the functionality of the crack profile from a straight-through notch to the semi-elliptical shape at fracture. Additionally, this may improve the data acquisition reliability at very short crack lengths by minimizing the number of crack nucleation sites.

References

[1] Roe, G. M. and Coffin, L. F., Jr., unpublished research, General Electric Corporate Research and Development Center, Schenectady, NY, 1978.
[2] Milne-Thompson, L. M., *Theoretical Hydrodynamics,* McMillan, 1960, pp. 506–511.
[3] Gangloff, R. P., *Fatigue of Engineering Materials and Structures,* Vol. 4, 1981, pp. 15–33.
[4] Van Stone, R. H. and Krueger, D. D., "Investigation of Direct Aged Inconel 718 Fatigue Behavior, Final Report," NAVAIR Contract N00019-82-C-0373, GE Aircraft Engines, Cincinnati, OH, 1984.
[5] Ileubaum, F. H., "Propagation Kinetics of Short Fatigue Cracks in Low Alloy Steels: Crack Closure and Fracture Morphology," Ph.D. dissertation, Northwest University, Evanston, IL, 1986.
[6] Shih, Yan-Shin and Chen, Jien-Jong, "Analysis of Fatigue Crack Growth on a Cracked Shaft," *International Journal of Fatigue,* Vol. 19, No. 6, 1997, pp. 477–485.

[7] Carpinteri, A. and Brighenti, R., "Part—Through Cracks in Round Bars Under Cyclic Combined Axial and Bending Loading," *International Journal of Fatigue,* Vol. 18, No. 1, 1996, pp. 33–39.

[8] Henkener, J. A. and Forman, R. G., "Fatigue Crack Growth and Fracture Toughness Properties of Several High Strength Bolt Materials," Lyndon B. Johnson Space Center, JSC-25480/LESC-29931, 1991, pp. 1–14.

[9] Carpinteri, A. and Brighenti, R., "Fatigue Propagation of Surface Flaws in Round Bars: A Three-Parameter Theoretical Model," *Fatigue Fracture Engineering Materials Structures,* Vol. 19, No. 12, 1986, pp. 1471–1480.

[10] Forman, R. G. and Shivakumar, V., "Growth Behavior of Surface Cracks in the Circumferential Plane of Solid and Hollow Cylinders," *Fracture Mechanics: Seventeenth Volume, ASTM STP 905,* American Society for Testing and Materials, 1986, pp. 59–74.

[11] James, L. A. and Mills, W. J., "Review and Synthesis of Stress Intensity Factor Solutions Applicable to Cracks in Bolts," *Engineering Fracture Mechanics,* Vol. 30, No. 5, 1988, pp. 641–654.

Louis Raymond[1]

Accelerated Small Specimen Test Method for Measuring the Fatigue Strength in the Failure Analysis of Fasteners

REFERENCE: Raymond, L., "Accelerated Small Specimen Test Method for Measuring the Fatigue Strength in the Failure Analysis of Fasteners," *Structural Integrity of Fasteners: Second Volume, ASTM STP 1391,* P. M. Toor, Ed., American Society for Testing and Materials, West Conshohocken, PA, 2000, pp. 192–203.

ABSTRACT: In conducting a failure analysis on a fastener, measuring fatigue strength or endurance limit is often the time-limiting factor in the analysis. The classical approach to measure the fatigue strength, which often takes as much as 3–4 months to 3–4 years using several machines, cannot be tolerated from an operational cost perspective, especially if the failure closes down production.

This paper presents an accelerated method of testing for the fatigue strength of a fastener based on measuring the threshold stress for fatigue crack initiation at the root of a thread. The method builds on the work of Prot and Corten in combination with a Rising Step Load™-fatigue testing protocol. Using this new approach, the fatigue strength can be measured in less than one week with one machine.

A case history is presented to illustrate the use of this new method in fatigue to quantify the effect of surface conditions at the root of the threads on the life of a large 3.5-in.-diameter (8.9 cm) bolt. Life analysis is based on a Goodman Diagram modified within the framework of fracture mechanics.

KEYWORDS: failure analysis, fatigue strength, threshold, endurance limit, fasteners, screws, bolt, life analysis

Conducting any failure analysis inherently demands that the results be obtained in the shortest possible time; therefore, the tests must be *accelerated.* Time to the manufacturer means large amounts of money lost during a production shutdown.

When dealing with fasteners, testing is inherently restricted to *small specimens.* Screws have finite size that can be as small as 0.16 in. (4 mm) in diameter. Bolts can be as large as 4-in. (100-mm) in diameter. The geometry of the threads is standardized. There is no limit on length. For a full-size bolt, the large loads required for testing become restrictive and small specimens cut from the bolt must be used.

Overload failures occur instantaneously with peak loads and are quickly verified with the scanning electron microscope (SEM) by the presence of dimples on the fracture surfaces. Failure of a fastener in service is generally related to time delayed fracture either by fatigue or environmentally induced, subcritical crack growth.

[1] Principal consultant, L. Raymond & Associates, Newport Beach, CA 92658.

Concept of Threshold

For both fatigue and environmentally induced subcritical crack growth, the design evolves around the concept of a threshold. Current design practice based on crack propagation references the threshold stress intensity for fatigue crack propagation as ΔK_{th}. In stress corrosion cracking, K_{Iscc} is the parameter most frequently referenced in designing against environmentally assisted subcritical crack propagation.

In classical *S-N* fatigue testing, the threshold alternating stress below which no fracture will occur is referenced to as the fatigue strength or endurance limit. Examining Fig. 1, it should be noted that this fatigue threshold alternating stress corresponds to the threshold stress for crack initiation, which is not to be confused with the threshold stress intensity for crack propagation.

In conducting a *failure analysis* to determine what factors reduced the projected design life of a fastener, the fatigue strength can be obtained by measuring the *threshold* stress for crack initiation. It is the separation betwen finite and infinite life. As the threshold is approached in the finite life region, the time it takes to initiate or nucleate a crack represents the largest portion of the total life as illustrated in Fig. 1 [*1*]. Once a crack is initiated, the clock on remaining life based on crack propagation begins to run.

Classic Approach

Classically, the threshold is measured by applying a constant load (alternating in fatigue and static in stress corrosion testing) at various stresses of decreasing magnitude on a series

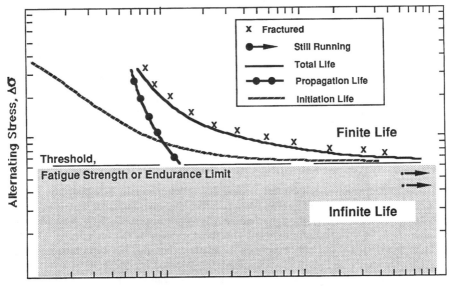

Number of Cycles to Failure, N

FIG. 1—*A conventional* S-N *curve (stress vs. number of cycles to fracture) can be separated into initiation life and propagation life (after Barsom and Rolfe [1]. The fatigue strength delineates the region of "infinite" life from "finite" life. The threshold stress for crack initiation is equivalent to the fatigue strength.*

of specimens of a specific geometry until a threshold stress is measured below which no fracture will occur. The practical difficulty with the classical approach from failure analysis considerations is that the runout times to establish these parameters are excessive. In stress corrosion testing of steels at 33 HRC-35 HRC, the U.S. Navy calls for a runout time of four to five years! The current ASTM Standard Test Method for Determining a Threshold Stress Intensity for Environment-Assisted Cracking of Metallic Materials under Constant Load (E 1681) calls for $\geq$10 000 h (>1 year) for steels @ <1000 MPa (170 ksi $\approx$ 38 HRC).

Classically, fatigue testing of steels consists of testing a group of 12 to 15 specimens, each under a different constant cyclic load until fracture occurs (see Fig. 1). The stress level at which the specimens do not fracture after $\geq$10^6 cycles, approximately defines the "fatigue strength" or "fatigue endurance limit," or the "threshold stress" below which no fatigue fracture will occur, or the region of "infinite life." One machine is required to test each specimen, and one runout test may take as long as four weeks. Therefore, measurement of the threshold fatigue stress would require several months on several machines.

Since the conventional test methods are not practical within the time constraints of most failure analyses, an *accelerated* method to measure the threshold in both fatigue and environmentally induced cracking is of utmost importance. With fasteners, the test specimens are limited by the size of the fastener. For a large bolt, testing must often be conducted by using *small specimens* cut from the bolt. These results must be scaled up within the framework of fracture mechanics to obtain threshold measurements for the large diameter bolt.

New Accelerated Approach

Over the last 20 years, work has been conducted to provide an accurate and accelerated test method for the measurement of the threshold stress in both stress corrosion cracking and fatigue. A method was developed on notched specimens for environmentally assisted subcritical crack growth that measured an "effective" threshold stress intensity for crack initiation [2,3] that was applied to fasteners. This test method was also successfully used on sharp cracked specimens to measure K_{Iscc} [4]. The same approach is now being used to provide and an accelerated approach to measurement of fatigue strength. As a result, the RSL method has been found to provide verifiable and reproducible measurements of the threshold stress intensity for both fatigue and environmentally-assisted subcritical crack growth on one machine in less than one week.

Environmentally Assisted Subcritical Crack Growth

The accelerated, small specimen test method that is commercially designated as the Rising Step Load™ (RSL™),[2] already exists for measuring the environmentally assisted cracking threshold of fasteners [5]. Well established standards are in place. ASTM Test Method for Measurement of Hydrogen Embrittlement Threshold in Steel by the Incremental Step Loading Technique (F 1624) not only describes the method, but also has been incorporated into many other standards such as ASTM Standard Test Method for Mechanical Hydrogen Embrittlement Evaluation of Plating Processes and Service Environments (F 519) and ASTM Test Method for Process Control Verification to Prevent Hydrogen Embrittlement in Plated or Coated Fasteners (F 1940).

---

[2] Rising Step Load™ (RSL™) is a trademark of Fracture Diagnostics, Inc., Denver, CO.

Influence of the Environment

Environmentally induced failures use many different abbreviations. On examination of these presumably different failure mechanisms, other than the type of environment, in steels they can be reduced down simply to differences in the sequence of application of the stress and the environment. Even classical internal hydrogen embrittlement (IHE) and stress corrosion cracking (SCC) or environmental hydrogen embrittlement (EHE) are caused by the same mechanism in high-strength steels used for fasteners. Since most fasteners are manufactured from steel plated with a metallic or other inorganic coating, these failure mechanisms represent a very large percentage of the total failures in fasteners.

All environmentally induced subcritical crack growth results from a combination of applied stress and environmental exposure. It is the critical combination of hydrogen concentration at a point of maximum local stress that initiates hydrogen-assisted subcritical crack growth. Classical hydrogen embrittlement occurs when the part is first exposed to an environment (acid cleaning, pickling, electrolytic deposition of plating) and then exposed to stress (unless a residual stress existed in the part before plating. Then the part in all likelihood would break in the plating bath).

By comparison, a part in service that has been produced with no residual hydrogen will first be stressed and then exposed to an aqueous service environment that produces hydrogen. A secondary consideration, and often the most important consideration, is the galvanic couple between the steel fastener and the sacrificial anodic coating. From considerations of accelerated electrochemical corrosion attack based on both potential and area differences, fasteners are generally plated to be cathodic to the structure that they are holding together. In the presence of moisture, the coating sacrificially corrodes, protecting exposed portions of the steel from corroding (rusting), but in so doing, generating nascent hydrogen at the cathodic surface of the steel fastener. Under stress, the nascent or atomic hydrogen is being quickly and efficiently absorbed by the steel fastener. Once the critical combination of hydrogen at the location of a maximum local stress is attained, hydrogen-induced subcritical crack growth is initiated.

Fatigue

An accelerated, small specimen test method did not exist for measuring the fatigue strength of fasteners, which is necessary to conduct a failure analysis in a reasonable time. Based on the author's review of the literature, the most promising approach to measure the fatigue strength in an accelerated manner is to take advantage of the Prot method [6]. He demonstrated that by using a progressively increasing loading rate until the specimen fractures, the fatigue strength or endurance limit could be estimated in less than one week, on one test machine. As shown by Fig. 1, since the crack propagation region represents a small percentage of the total life, time to fracture measurements were used to approximate the time for crack initiation, which is a more precise interpretation of a fatigue strength.

Prot Method—Prot used smooth, rotating beam specimens ($R = -1$) on a low-alloy, a mild, and a high-carbon steel. He plotted the fracture stress as a function of the square root of the loading rate and showed that the experimental points practically fall in a straight line whose intersection with the ordinate axis is the endurance limit. He did not compare the results with the endurance limit measured by the conventional constant-stress-amplitude method of measuring the fatigue strength.

Appraisal—In 1954, Corten et al. [7] did an appraisal of the Prot method of fatigue testing. They used the same test procedure as Prot except that they also included a comparison of the results of the Prot method with the results from the conventional constant-stress-amplitude method of measuring the threshold or fatigue strength, including both smooth and notched round bar specimens at $R = -1$.

Corten et al. concluded that for rapid estimates of the fatigue strength, the use of the square root of the loading rate gives satisfactory results if a smooth curve is drawn through the data. They also concluded that fatigue strength determined by progressive loading of notched specimens was in close agreement with the values obtained by the constant-stress-amplitude method, which indicated that the Prot method may be applicable to members of any shape, including full-size structural elements like fasteners.

Corten pointed out that the Prot method of fatigue testing gives information only about the fatigue threshold and that it is most promising for ferrous materials with a well-defined endurance limit and not for aluminum alloys without a well-defined limit.

The use of lower loading rates was also recommended to prevent plastic bending after the development of a small crack in the rotating bend test. Fast loading rates can exceed the yield strength too quickly, causing permanent plastic bending of the specimen when tested under load control.

"Coaxing" was also evaluated and found to be nonexistent in the higher-strength, low-alloy steels as illustrated by Prot's data shown in Table 1. Only in a low-strength, ingot iron, which was specifically processed to be most responsive to "coaxing" during the progress of the repeated load test, were Corten et al. able to demonstrate "coaxing" with the Prot method. The special processing consisted of quenching the ingot iron in water to retain as much of the carbon and nitrogen in solid solution as possible instead of using it in the normal hot-worked condition. As anticipated, the maximum difference was obtained and found to be 5 ksi (35 MPa).

New Approach—The work of Corten et al. established the usefulness of the Prot method as an accelerated test method of measuring the fatigue strength of notched specimens or structural elements such as steel fasteners. To adapt this information to support the failure analysis of fasteners, the test procedure had to be modified and the technology had to be updated to predict the life of a fastener as defined by the Goodman diagram within the framework of fracture mechanics.

The new approach consists of combining the work of Prot and Corten et al. with the RSL-fatigue loading protocol to accurately define the onset of crack initiation in an accelerated fashion. The specific contributions of this paper to attain this goal are: (1) the use of a Rising Step Load (RSL) profile at a constant R-ratio of 0.1 instead of a continuously increasing reverse bending load at an R-ratio of -1 in order to accurately measure the onset of crack initiation; (2) the use of four-point bend loading and fastener specimens or square bar specimens machined from fasteners that duplicate thread conditions of a fastener instead of a smooth specimen; (3) the use of displacement control to measure crack initiation by a load drop to terminate the test instead of fracture of the specimen, and finally; (4) performing the

TABLE 1—*Comparison of endurance limit obtained by the Prot method and the conventional method.*

YS/TS[a]	Steel	Hardness	Notched[b]	Smooth[b]
242/271	14B50	51HRC	(37.0/39.0)	(60.0/61.0)
92/113	SAE 2340	99HRB	(39.2/38.0)	(71.7/69.0)
33/55	Ingot Iron[c]	54HRB	(.../...)	(28.8/23.0)

[a] YS/TS = Yield/Tensile Strength, ksi.
[b] Threshold Stress, ksi (Prot/Conventional).
[c] Ingot iron specially processed to maximize the response to "coaxing."
NOTE—Multiply ksi by 6.895 to convert to MPa.

analysis in the framework of fracture mechanics by using an "effective" stress intensity, which is further used to develop a fastener-modified Goodman Diagram for life analysis. This approach is especially effective in analyzing for differences in fasteners with surface treatments such as with rolled threads, case-hardened threads, or plated or coated threads.

Advantages—The specific advantages of each of these modifications are detailed as follows:

1. The use a Rising Step Load (RSL) profile at a constant R-ratio:

• Uses a modern, programmable, closed-loop, hydraulic, computer-controlled fatigue machine under displacement control, instead of the relatively primitive test methods employed by Prot and Corten et al. where they used a rotating beam test with a hanging bucket being filled with water at different rates.

• In comparison with a continuously increasing loading rate used by Prot and Corten et al., the incremental step-load profile allows accurate detection of the threshold for the onset of crack initiation.

• The ability to vary the R-ratio allows for the determination of a Goodman Diagram.

2. The use of four-point bend loading and fastener specimens that duplicate thread condition:

• Four-point loading produces a constant moment that can be used to calculate the stress anywhere along the threaded or irregular cross section of the specimen, permitting calculation of the maximum stress at the root of the thread.

• Using an actual fastener of a specimen cut from a large bolt that retains the thread profile on the tensile side of the specimen has the obvious advantage of representing the actual surface conditions of the thread surface.

• As a result, a study can be conducted on the effect of the surface conditions of an actual fastener on the fatigue life and subsequent analysis with the Goodman diagram.

3. The use of crack initiation to terminate the test:

• Conducting the test under displacement control maintains the load constant until a crack initiates that changes the compliance of the system. This change manifests itself as a load drop that is used to define crack initiation; thus, crack initiation is defined as a 5% drop in load.

• Although near-threshold crack growth represents a small percentage of the total life, it is removed from the analysis by using the crack initiation criterion, which reduces the total time of the test.

4. Performing the analysis in the framework of fracture mechanics by using an "effective" stress intensity:

• Consistent with one of the main advantages of fracture mechanics, test results on a small specimen can be used to predict the structural behavior of a large bolt.

• For small screws, the head is removed and the threaded portion is inserted into an adapter on each end, resulting in a dogbone-shaped specimen.

Testing Protocol

Test Specimen

In fatigue testing fasteners, if at all possible, the specimens should be exemplars, taken from the same lot, if possible. In this way, processing is removed as a variable. For a failure analysis, if any accuracy has to be sacrificed, it should be done in meeting the rigorous requirements of specimen geometry rather than altering the surface conditions in specimen preparation. Therefore, the specimens should be made from the actual fasteners.

For small-diameter fasteners or screws, the entire fastener can be used. Cut off the head and drill and tap both ends of square bar adapters to produce a "dogbone" type specimen as shown in Fig. 2a.

For large-diameter fasteners or bolts, a square piece can be machined with three sides remaining flat and perpendicular, and the fourth side retaining the threads from the large-diameter bolt. Figure 2b shows the actual dimensions of a specimen cut from a 3 3/4-16 UNJ bolt.

Loading Method

As noted in Fig. 2a or Fig. 2b, the loading method is four-point bending. The moment (M) in either case is given by $P \Delta / 2$, where Δ is 0.5 in. (1.27 cm) and P is the load measured with the closed-loop hydraulic test machine.

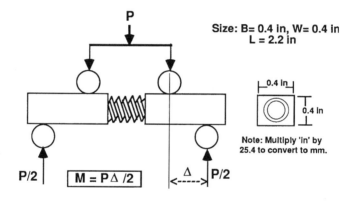

a — Adapters for testing small screws in 4-point bending.

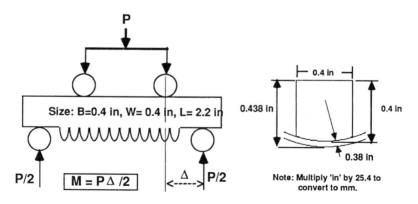

b — Adapters for testing in 4-point bending, square bar specimens

machined from large diameter bolts.

FIG. 2—*Specimens and adapters used to measure the onset of cracking with the four-point loading method under displacement control. (a) Adapters for testing small screws in four-point bending. (b) Adapters for testing in four-point bending, square bar specimens machined from large diameter bolts.*

Stress Calculations

For small screws (Fig. 2a), the stress is given by $32\,M/\pi\,d^3$, where d is the minor diameter. For the square bar specimens cut from a large-diameter bolt (Fig. 2b), the stress is given by $6\,M/W^3$, where W is the width, which equals the thickness, $B = 0.4$ in. (10 mm).

Stress Intensity Calculations

For small screws (Fig. 2a), the stress intensity is given by

$$K_{\text{Ieff}} = \sigma\,\sqrt{D}\,f(d/D) \tag{1}$$

where D is the major diameter and $f(d/D)$ can be found in Section 27.3 in Tada et al. [8]. For the square bar specimens cut from a large-diameter bolt (Fig. 2b), the stress is given by

$$K_{\text{Ieff}} = \sigma\,\sqrt{\pi a}\,f(a/W) \tag{2}$$

where a is the thread depth and $f(a/W)$ can be found in Section 2.13 in Tada et al. [8].

K_{Ieff} is defined as an "effective" stress intensity parameter used for comparative purposes and is not a stress intensity parameter in an absolute sense, because of the radius of the threads instead of a sharp crack, but, it does serve the need of identifying the root cause of a fastener fatigue failure.

Loading Spectrum

The step-load profile is shown in Fig. 3. Each load block is maintained at a stress ratio of minimum to maximum stress of $R = 0.1$. The width of each block or time increment is either 1, 2.25, or 4 h. The height or load is based on an estimate of attaining about one-half the tensile strength in about 20 h, which in the case of a 12 chromium steel at 125/150 ksi (860/1030 MPa) yield/tensile calculated to be 160 lb (7.1 kN).

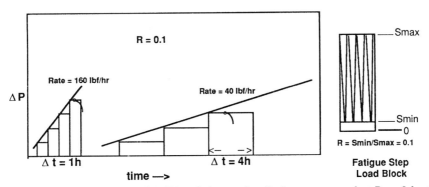

FIG. 3—*Cyclic Rising Step Load (RSL) technique under displacement control at* R = 0.1 *at two different rates to determine the onset of cracking detected by a load drop. Note: Multiply 1bf by 4.45 to convert to N.*

Loading Rate

The fastest loading rate was 160 lb/h (712 N/h) and the slowest loading rate was 40 lb/h (178 N/h). As an example, the first step ran for 1 h between 16 and 160 lb and then the load was increased to run between 32 and 320 lb/h, 48/480 lb/h, etc. until a crack was detected. The test was conducted at 20 Hz.

Since load and displacement are proportional, the test was conducted under displacement control, while monitoring the load. A load drop would be an indication of crack initiation, because of the corresponding change in compliance.

Crack Detection

The test is conducted under displacement control, which for the square bar approximates a constant stress intensity that is proportional to the load. Once a crack initiates, the load begins to drop. A criterion of a 5% load drop is used to define the presence of a crack, which corresponds to about 0.025 in. (0.6 mm).

Data Reduction

Figure 4 is an example of the method of plotting the data to obtain the threshold fatigue load for the specimen. Based on a relative scale, the fastest rate is unity and the square root of the slowest rate is two (2). The intermediate rate is the square root of 2.25 or 1.5. The total time to measure the threshold load on one machine is about one week or 1, 2, and 4 days for 1 h, 2.25 h, and 4 h load blocks, respectively.

Once the load *range* is measured from the modified Prot test, the alternating stress intensity or *amplitude* can be calculated from Eq 2 realizing that the amplitude is one-half the stress range. The results of these calculations are summarized in Table 2.

Data Analysis

Once the results are in terms of the stress intensity, a fracture mechanics-modified Goodman Diagram may be constructed as shown in Fig. 5. Once in terms of a stress intensity parameter, the axes can be converted to conventional Goodman diagram in terms of stress

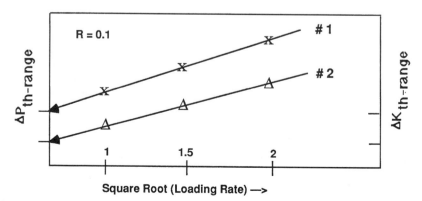

FIG. 4—*Extrapolation method of determining the threshold fatigue load range* ($\Delta P_{th\text{-}range}$) *with only three specimens cyclically loaded under displacement control at* R = 0.1.

TABLE 2—*Threshold loads (kips) at* R = *0.1 of specimen in Fig.* 2a *tested in accordance with the RSL-fatigue testing protocol and corresponding calculations of the alternating stress intensity from Eq 2.*

Thread	ΔP_{th}-range, kips	ΔK_{th}-alt, ksi $\sqrt{\text{in.}}$
Machined	3.3	11.6
Rolled	4.0	14.0
Rolled and Plated	1.7	6.0

NOTE—Multiply kips by 4.45 to convert to kN and ksi $\sqrt{\text{in.}}$ by 1.1 to convert to MPa $\sqrt{\text{m}}$.

amplitude and mean stress for a range of stress ratios by using Eq 1 for the 3 3/4-16 UNJ bolt. This conversion to stress is shown in Fig. 5.

Failure Analysis

In Fig. 5, different zones of "infinite life" are delineated, depending on how the threads were processed. The largest zone is the one with the rolled threads that show an improvement over the machined threads. Of significance is the degradation from plating as illustrated by the smallest "infinite life" zone.

The design stress was intended to operate below the threshold fatigue strength line for either machined or rolled threads. The unanticipated result that is the cause of the failure is the significant drop in the fatigue strength due to plating.

To add to the credibility of the results, the stress values of the threshold or fatigue strength at $R = -1$ from Fig. 5 were compared with the data of Almon and Black [9] shown in Fig.

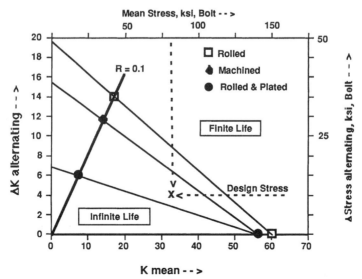

FIG. 5—*Fracture Mechanics Modified Goodman Fatigue Strength Diagram used to calculate the fatigue stress limits for a 3 3/4-16 bolt as a function of processing details of the threads.*

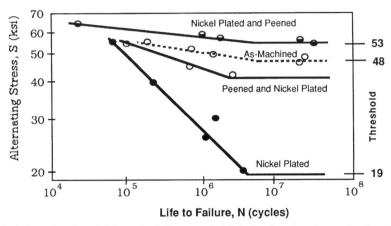

FIG. 6—*Effect of surface finish on the* S-N *curve of nickel plated steel at* R = −1. *After Almon and Black [9].*

6. As can be observed in Table 3, both sets of results show comparable degradation from plating. It should also be noted that the results of Almon and Black were conducted using conventional fatigue testing protocol as compared with those of the present program using the accelerated, small-specimen testing protocol.

Conclusions

A new, accelerated, fatigue testing method was identified that can be used as a tool for failure analysis of fasteners. The fatigue strength as measured by the threshold stress intensity for fatigue crack initiation can be measured in one week with one machine as compared with the classical long-time test that is impractical for failure analysis.

The specimens are made from exemplar fasteners obtained from the same lot as the failed fastener, which eliminates material processing as a variable. Screws can be bested directly as manufactured. Specimens that retain the thread on one surface can be machined from large bolts. Different conditions of the surface of the threads can be studied to isolate their contribution to the failure process. The analysis is conducted on small specimens using an "effective" stress intensity, which permits the results to be applied to different-size fasteners with the same root radius.

The test method has been demonstrated on a specific class of steels and is most promising for ferrous materials with a well-defined endurance limit.

TABLE 3—*Corresponding alternating threshold stress (ksi) at* R = −1.0 *from Fig. 5 as compared to the alternating threshold stress for a nickel plated steel tested by the conventional method [9].*

Thread	$\Delta\sigma_{alt}$	$\Delta\sigma_{alt}$	Surface
Machined	39.0	48	As-Machined
Ni-Cd Plated	17.5	19	Ni-Plated

NOTE—Multiply ksi by 6.895 to convert to MPa.

References

[*1*] Rolfe, S. T. and Barsom, J. M., *Fatigue and Fracture Control in Structures, Application of Fracture Mechanics,* Prentice-Hall, Englewood Cliffs, NJ, 1977, p. 209.

[2] Raymond, L., "Application of a Small Specimen Test Method to Measure the Subcritical Cracking Resistance in HY-130 Steel Weldments," under contract No. DTNSRDC-SME-CR-09-82 entitled, "High-Strength Steel Weldment Subcritical Cracking Program," under Navy Contract N00167-81-C-0100, LRA Labs (previously named METTEK), Final Report No. 210123, March 1982.

[*3*] Raymond, L. and Crumly, W. R., "Accelerated, Low-Cost Test Method for Measuring the Susceptibility of HY-Steels to Hydrogen Embrittlement," *Proceedings of the 1st International Conference on Current Solutions to Hydrogen Problems in Steel,* American Society for Metals, Metals Park, OH, Nov. 1982, pp. 477–480.

[*4*] Raymond, L., "Accelerated Stress Corrosion Cracking Screening Test Method for HY-130 Steel," under SBIR Phase I and Phase II NAVSEA contract No. N00024-89-C-3833, LRA Labs, Final Report No. NAVSEA 80058, Dec. 1989 and Dec. 1993, respectively.

[5] Raymond, L., "The Susceptibility of Fasteners to Hydrogen Embrittlement and Stress Corrosion Cracking," *Handbook of Bolts and Bolted Joints,* Chapter 39, Marcel Decker, Inc., New York, 1998, pp. 732–756.

[6] Prot, E. M., "Fatigue Testing Under Progressive Loading: A New Technique for Testing Materials," *Revue de Metallurgie,* Vol. XLV, No. 12, p. 481 (1948), English translation by E. J. Ward, WADC Technical Report 52-148, Wright Air Development Center, Sept. 1952.

[7] Corten, H. T., Dimoff, T., and Dolan, T. J., "An Appraisal of the Prot Method of Fatigue Testing," American Society for Testing and Materials, preprint #69, 1964.

[8] Tada, H., Paris, P., and Irwin, G., *The Stress Analysis of Cracks Handbook,* Paris Productions, Inc., St. Louis, MO, 1985. Section 2.13 and Section 27.3.

[9] Alman, J. O. and Black, P. H., *Residual Stresses and Fatigue in Metals,* McGraw-Hill, New York, 1963.

Harold S. Reemsnyder[1]

Fracture Mechanics of Mechanically Fastened Joints—A Bibliography

REFERENCE: Reemsnyder, H. S., **"Fracture Mechanics of Mechanically Fastened Joints—A Bibliography,"** *Structural Integrity of Fasteners: Second Volume, ASTM STP 1391,* P. M. Toor, Ed., American Society for Testing and Materials, West Conshohocken, PA, 2000, pp. 204–214.

ABSTRACT: The application of fracture mechanics to mechanical fasteners requires stress-intensity-factor solutions for cracks at fastener holes and in threaded, round bars. This bibliography lists references that present stress-intensity-factor solutions for cracks at circular holes and a wide range of cylindrical, solid and hollow, round bars, unnotched, and circumferentially notched, subjected to axial loads, bending. Notched cases include shoulder fillets, multiple thread-like circumferential projections, and thread-like single circumferential grooves in round bars. In these cases, the projections and grooves were not helical. Instead, the planes of the projections and grooves were perpendicular to the axis of the round bar. The geometries of cracks in round bars include circumferential cracks and cracks initiated at a point on the surface of the cylinder. The latter crack geometries include straight, curved, semicircular, semi-elliptical, and sickle-shaped crack fronts. A few references treat high-strength bolts.

KEYWORDS: stress intensity factors, fasteners, bolting, weight functions, fracture mechanics, cracks, crack shape, crack growth, compliance

Nomenclature

a	Crack depth
b	Half-crack length
D	Diameter of a circular bar
K_t	Elastic stress concentration factor
K_I	Mode I stress intensity factor
K_{II}	Mode II stress intensity factor
Y	Correction factor in $K_I = S \cdot \sqrt{\pi a} \cdot Y$, a function of loading condition and local and overall geometry

The description of fatigue crack growth and fracture in mechanically fastened joints by stress-intensity-factor concepts is complicated by:

- the statically indeterminate nature of the stress distribution in the plies along, and across, the hole array,
- the change in stress distribution due to ply stiffness changes with growth, and

[1] Senior research consultant, Bethlehem Steel Corporation, Homer Research Laboratories, Bethlehem, PA 18016-7699.

- the influence of hole proximity to crack tip on stress intensity factor, i.e., notch shadow effect.

These complications are addressed in the stress analysis of mechanically fastened joints and will not be mentioned herein. Instead, only references discussing the applications of fracture mechanics to cracks at fastener, i.e., circular, holes and to cracks in cylindrical bars will be listed.

Cracks at Circular Holes

Stress intensity factors have been estimated for three geometries for cracks in the joined plies at circular holes:

- through-thickness cracks,
- semi-elliptical surface cracks in the bore of the hole, and
- quarter-elliptical or quarter-circular corner cracks at the intersection of the bore with the surface of the ply.

For brevity, the three geometries will be called, respectively, *radial, surface* and *corner* cracks.

The planes of these cracks are perpendicular to the surface of the ply and contain the axis of the circular hole. Also, the crack may be single or double, i.e., lying on, respectively, one side or both sides of the hole.

Fatigue crack propagation from open rivet holes is studied in Refs *1* and *2*. Reference *1* also discusses pin-loaded holes. Radial cracks at pin-loaded holes and corner cracks at open holes are studied in, respectively, Refs *3* and *4*. Reference *5* reviews the state of the art (to 1975) for dealing with cracks at holes in engineering structures and discusses effects of fasteners, holes in reinforced structures, and the retardation and arrest capabilities of holes.

Finite-element analysis has been used to develop the stress intensity factor solutions for corner cracks at open holes in tension [6,7], corner cracks at open holes in bending, wedge-loaded and pin-loaded holes [6], and radial cracks at open holes under uniaxial stress [7]. Mixed-mode stress intensity factors have been developed for single and double radially cracked holes in bolted joints using *finite-element analysis* [8].

Stress intensity factor solutions for radial cracks at open holes under biaxial stress [9], surface cracks in the bore of a hole subjected to crack-face pressure, biaxial loading, wedge-loading, and pin-loading [10], and for surface and corner-cracked fastener holes [11] have been developed by the *weight function* method.

The *superposition method* has been used to develop stress intensity factors for corner cracks at an open hole under remote tension and crack-face pressure loading [12] both radial and corner cracks emanating from both open and loaded holes in finite width plates, lugs, and multi-fastener joints [13].

The *slice synthesis method* was used to predict fatigue crack growth of corner cracks and the residual stress concomitant with crack-tip yielding [14]. The method was then used to model crack retardation and acceleration due to these residual stresses in the presence of variable amplitude loading.

An approximate solution for the stress intensity factor for a radial crack in the vicinity of a circular hole in uniaxial tension was developed through the *compounding method* [15]. Using the method of Ref *15*, approximate stress intensity factors were developed for a radial crack at a single hole and, a periodic array of radially cracked holes, a periodic array of pressurized holes, a pin-loaded hole with two radial cracks [16], and two radial cracks at a hole in a tension-loaded strip [17].

Approximate methods have been used to develop stress intensity factors for single and double radial cracks at a hole and surface cracks in the hole-bore in uniaxial tension and containing an interference plug, and both radially cracked and surface-cracked holes subjected to a concentrated load [*18*]. Also, approximate stress intensity factors have been estimated for single and double radial cracks at a single hole in an array of holes [*19*].

Stress intensity factors can be *estimated empirically* by growing and measuring a fatigue crack, with periodic marking, in a particular geometry and combining these measurements with those from tests on conventional fatigue crack growth specimens. This technique has been used to estimate the stress intensity factor for double corner-cracked holes [*20–22*] and single corner-cracked holes [*22*] in tension.

Crack growth in riveted and friction-bolted structural joints is discussed in Ref *23*.

Life Extension of Cracked Holes

The use of fracture mechanics in aircraft joints with open, cold-worked, and pin-loaded holes, and with interference fit fasteners to retard fatigue crack growth is described in Refs *24* and *25*. (Reference *24* contains an extensive bibliography.) The use of high-strength bolts to retard or arrest crack growth in structures is described in Refs *23* and *26*. Both finite-element analyses and tests showed that bonded sleeves and/or bonded patches reduced the fatigue crack growth rate for two corner cracks at an open holes [*27*].

Compendia of Stress Intensity Factors for Cracked Holes

Stress intensity factor solutions for various cases of cracked holes are available in published compendia: Rooke and Cartwright [*28*], Tada, Paris and Irwin [*29*], and Murakami [*30*].

Solutions for radial cracks at an open hole under remote tension and internal pressure are presented in Refs *28, 29,* and *30*. Solutions are presented for radial cracks at an open hole under biaxial stress in Refs *28* and *29* and for radial cracks at a pin-loaded hole in Ref *30*. Solutions for a crack growing toward an open hole are presented in Refs *28, 29,* and *30* and for cracks between open holes in Refs *29* and *30*.

Cracks in Round Bars

The application of fracture mechanics to the fasteners themselves requires stress intensity factor solutions for cracks in round bars. Stress intensity factors have been estimated for circumferentially cracked bars and bars with cracks growing from one point on the surface of the bar. The crack planes of both geometries are perpendicular to the long axis of the bar. The crack shapes are either semicircular, semi-elliptical, straight, or sickle-shaped.[2] Both unnoticed bars and circumferentially notched bars have been studied. The latter have been used to simulate the effect of threads and are either:

- circumferential grooves to simulate the thread root, or
- circumferential projections (two or four) to simulate the thread.

In all cases, the notches were in a plane perpendicular to long axis of the round bar and did not replicate the helical path of actual threads. In most cases, the notches acted as stress

---

[2] In a sickle-shaped crack, the midpoint of the crack front lags the ends of the crack front.

raisers on remotely applied stresses. Only a few references included the thread load in the determination of stress intensity factor.

Two dimensional finite-element analysis has been used to determine stress intensity factors in unnotched and notched bars as given in Table 1.

Two-dimensional, axisymmetric finite-element analysis was used to estimate the Mode I and Mode II stress intensity factors, respectively, K_I and K_{II}, for cracks in a bolt-nut thread interaction (circumferential, not helical, notch) [43]. In these analyses, K_I was much greater than K_{II}.

Two-dimensional, axisymmetric finite-element analysis was used to model various threaded drillstring connections under axial, bending and torsion loads for input to subsequent fracture mechanics analyses [44].

A high-strength aircraft bolt was modeled by a cylindrical bar with four thread-like circumferential (not helical) projections through *three-dimensional finite-element analysis* [45]. The analysis was performed for four different depths of a semicircular crack located in the simulated thread root. Two load cases were studied—axial load and axial load plus thread load. It was shown that K_I for the case of axial load plus thread load was greater than K_I for axial load only, Fig. 1*a*. Also, it was shown that the stress intensity factor K_I increases from the midpoint of the crack-front to the intersection of the front with the surface of the bar, Fig. 1*b*.

Three-dimensional finite-element analysis has modeled semicircular and semielliptical cracks in a tension-loaded unnotched bar (with K_I estimated from the *J*-integral) [46] and an unnotched bar in tension and bending with semicircular and straight cracks [47].

Weight functions for a sickle-shaped crack in an unnotched bar under varying stresses were derived from a finite-element analysis of the same geometry subjected to a uniform tension [48].

TABLE 1—*Stress intensity factors.*

Crack Shape	Bar	Loading	Remarks	Refs
Straight	unnotched grooved,	tension, bending	circumferential	31
	unnotched	lateral compression to simulate shrink-fit		
Straight	unnotched	tension	measured compliance	32
Circular	unnotched	tension, pure bending		33
Straight	unnotched	pure bending		34
Circular	unnotched	tension, bending		35
Elliptical	two projections	tension, bending	circumferential	36
Elliptical	unnotched	tension, bending		37
Surface	shoulder fillet	pure bending	circumferential, corroborated with fatigue crack growth tests	38
Elliptical	three projections	tension, bending, residual and tightening stress	circumferential	39
Circular	notched	loaded by nut	circumferential	40
Circular, straight	unnotched	tension, pure bending		41
Surface	cut fillet, rolled filled	shear tension	under bolt head	42

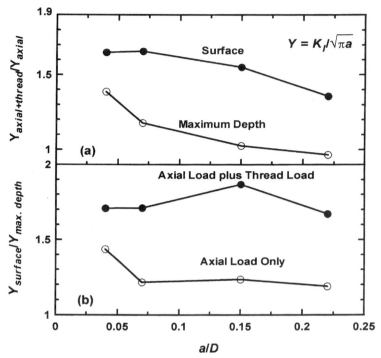

FIG. 1—*Three-dimensional finite-element analysis of a M14 × 1.5 bolt [45].*

Weight functions for circular cracks of varying radius in an unnotched bar pure bending were derived from a finite-element analysis of the same geometry subjected to a uniform tension [49].

The *weight function method* was used to develop stress intensity factors for a circumferentially grooved (not helical) bar in [50,51]. In Ref 51, the root radius of the groove was varied to simulate various thread geometries.

The stress intensity factor computed through the *boundary integral method* for a semi-elliptical crack in an unnotched bar, tension or bending, was corroborated by compliance measurements [52].

The influence function method was used to determine the stress intensity factor of a semi-elliptical crack in a tension-loaded, circumferentially grooved bar [53]. The thread was introduced as a two-dimensional stress concentration factor K_t.

Stress intensity factors have been determined empirically for semicircular cracks in a tension-load, unnotched bar [54] and a circumferentially notched, tension-loaded bar [55].

Experiments

Stress intensity factors for a semi-elliptical crack in an unnotched bar under four-point bending were estimated from a combination of three-dimensional photoelasticity and crack growth measurements on the subject geometry [56].

Fatigue crack growth data (including periodic cyclic marking) from tests on tension-loaded high-strength bolts were combined with fatigue crack growth data from conventional test specimens to estimate the stress intensity factor of a cracked bolt [57].

Forman and Shivakumar measured the cyclic growth of circular cracks in solid, unnotched bars and elliptical cracks (both external and internal) in hollow, unnotched cylinders under tension and bending [58]. The stages of growth were established by periodic heat tinting.

The stress intensity factors for a straight crack in solid and hollow, unnotched bars in a three-point bending were determined experimentally from compliance measurements [59–61].

The cyclic growth of elliptical and circular cracks in a bolt loaded through the nut has been measured experimentally [62].

State of the Art Reviews

Daoud and Cartwright assembled results for curved and straight cracks in unnotched bars subjected to tension and bending in 1986 [63].

In 1988, James and Mills reviewed the literature presenting stress intensity for solutions for straight and semicircular cracks in both unnotched and notched bars [64].[3] The notched bar geometries were the single circumferential groove representing a thread root [53] and two circumferential projections representing two adjacent threads [36,54]. (Neither geometry modeled the thread helix.) The authors synthesized the three solutions (1) for notched tension, (2) semicircular crack in an unnotched bar, and (3) straight crack in an unnotched bar to develop a single relation for stress intensity factor K_I versus the ratio of crack depth to bar diameter a/D for a threaded bar in tension, Fig. 2. A similar synthesis was used to develop a K_I-a/D relation for a threaded bar in bending, Fig. 2. In the latter case, insufficient data were available to include notched bending.

Liu, in 1993, reviewed circular, elliptical, and straight cracks in unnotched and notched bars loaded in tension and bending [65].

In 1995, Liu reviewed circular, elliptical, and straight cracks in unnotched and notched bars loaded in tension [66]. He concluded:

- crack fronts are circular, not elliptical, and the radius of the crack front increases with crack depth approaching a straight crack front,
- the Forman-Shivakumar equation [58] for cracked, unnotched bars correlates well with test data for unnotched bars,
- the James-Mills synthesized equation for notched bars [64] is confirmed by subsequent finite-element analyses but lies on the upper bound to test data for notched bars,
- a single Y-curve where $Y = K_I/S \cdot \sqrt{\pi a}$ is inadequate for all thread geometries.

The James-Mills synthesized equations [64] for threaded (i.e., notched) bars and the Forman-Shivakumar equations [58] for unnotched bars are shown in Fig. 2 along with Liu's modification of the James-Mills equation for a threaded bar under tension [64].

Liu's conclusion [66] that the James-Mills synthesized equation for notched bars [64] is confirmed by subsequent finite-element analyses is illustrated in Fig. 3 where the analytical results of Torobio et al. [39] and Reibaldi and Eiden [45] are plotted. The James-Mills synthesized equation for threaded fasteners in tension agrees well with both investigators for small, semicircular cracks ($a/b = 1.0$).

Also, Liu's conclusion [66] that small-crack fronts are circular, not elliptical, and the radius of the crack front increases with crack depth approaching a straight crack front is illustrated in Fig. 3. As the crack grows larger, i.e., $a/D \to 1$, the portion of the James-Mills synthesized

---

[3] Reference *64* lists references (primarily German and Japanese) not included herein.

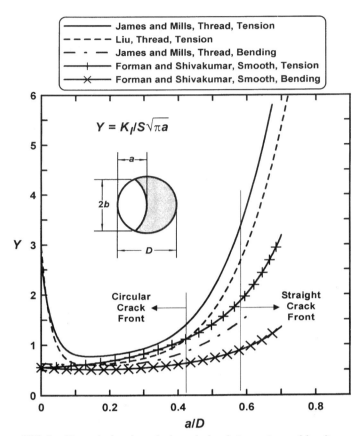

FIG. 2—*Unnotched and notched cracked rods in tension and bending.*

equation that is based on a straight crack front model is corroborated by the model of Torobio et al. [*39*] as $a/b \rightarrow 0$, i.e., the shape of the crack front approaches a straight line.

The reader is reminded that most of the so-called threaded bar cases actually treat a circumferentially notched bar (with either groove or thread-like projections) loaded by a remote tension. At short crack depths, the James-Mills synthesized equation agrees with the model of Reibaldi and Eiden where the remote tension was accomplished by thread loading [*45*], Fig. 3.

Compendia of Stress Intensity Factors for Cracked Round Bars

Stress intensity factor solutions for various cases of cracked round bars are available in published compendia: Tada, Paris and Irwin [*29*], Murakami [*30*], and Sih [*67*].

Solutions for circumferentially cracked, unnotched round bars in tension are presented in Refs *29, 30* and *67*, and for tension, bending and torsion in Ref *29*. Solutions for straight, semicircular and semi-elliptical cracks in an unnotched, round bar under tension and bending are presented in Ref *30*.

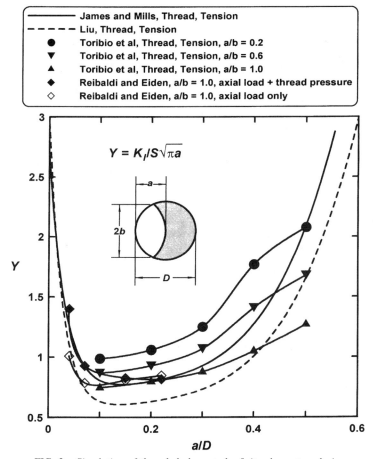

FIG. 3—*Simulation of threaded elements by finite-element analysis.*

Experiments

The electric potential method was used to monitor the formation and growth of a surface crack in an hour-glass-shaped specimen [68]. This study used a stress intensity factor developed previously [69].

The detection and measurement of the shape and growth of cracks in threaded elements using the A.C. potential method have been discussed in Refs 70 and 71.

References

[1] Figge, I. E. and Newman, J. C., Jr., "Fatigue Crack Propagation in Structures with Simulated Rivet Forces," *Fatigue Crack Propagation, ASTM STP 415,* American Society for Testing and Materials, West Conshohocken, PA, 1967, pp. 71–93.

[2] Broek, D. and Vlieger, H., "Cracks Emanating from Holes in Plane Stress," *International Journal of Fracture Mechanics,* Vol. 8, 1972, pp. 353–356.

[3] Cartwright, D. J. and Ratcliffe, G. A., "Strain Energy Release Rate for Radial Cracks Emanating from a Pin Loaded Hole," *International Journal of Fracture Mechanics,* Vol. 8, 1972, pp. 175–181.

[4] Liu, A. F., "Stress Intensity Factor for a Corner Flaw," *Engineering Fracture Mechanics,* Vol. 4, 1972, pp. 175–179.

[5] Broek, D., "Cracks at Structural Holes," MCIC-75-25, Metals and Ceramics Information Center, Columbus, Ohio, March 1975.

[6] Raju, I. S. and Newman, J. C., Jr., "Stress Intensity Factors for Corner Cracks at the Edge of a Hole," *NASA Technical Memorandum 78728,* June 1978.

[7] Kullgren, T. E., Smith, F. W., and Ganong, G. P., "Quarter Elliptical Cracks Emanating from Holes in Plates," *Journal of Engineering Materials and Technology,* Vol. 100, 1978, pp. 144–149.

[8] Ju, S. H., "Stress Intensity Factors for Cracks in Bolted Joints," *International Journal of Fracture,* Vol. 84, 1997, pp. 129–141.

[9] Petroski, H. J. and Achenbach, J. D., "Computation of the Weight Function from a Stress Intensity Factor," *Engineering Fracture Mechanics,* Vol. 10, 1978, pp. 257–266.

[10] Zhao, W., Wu, X. R., and Yan, M. G., "Weight Function Method for Three-Dimensional Crack Problems—II. Application to Surface Cracks at a Hole in Finite Thickness Plates Under Stress Gradients," *Engineering Fracture Mechanics,* Vol. 34, No. 3, 1985, pp. 609–624.

[11] Zhao, W. and Atluri, S. N., "Stress Intensity Factor for Surface and Corner Cracked Fastener Holes by the Weight Function Method," *Structural Integrity of Fasteners, ASTM STP 1236,* American Society for Testing and Materials, West Conshohocken, PA, 1995, pp. 95–107.

[12] Grandt, A. F., Jr. and Kullgren, T. F., "Stress Intensity Factors for Corner Cracked Holes under General Loading Conditions," *Journal of Engineering Materials and Technology,* Vol. 103, 1981, pp. 171–176.

[13] Ball, D. L., "The Development of Mode I, Linear-Elastic Stress Intensity Factor Solutions for Cracks in Mechanically Fastened Joints," *Engineering Fracture Mechanics,* Vol. 27, No. 7, 1987, pp. 653–681.

[14] Fujimoto, W. T. and Saff, C. R., "A Multidimensional-Crack-Growth Prediction Methodology for Flaws Originating at Fastener Holes," *Experimental Mechanics,* Vol. 19, 1982, pp. 139–146.

[15] Cartwright, D. J. and Rooke, D. P., "Approximate Stress Intensity Factors Compounded from Known Solutions," *Engineering Fracture Mechanics,* Vol. 6, 1974, pp. 563–571.

[16] Rooke, D. P., "Compounded Stress Intensity Factors for Cracks at Fastener Holes," *Engineering Fracture Mechanics,* Vol. 19, No. 2, 1984, pp. 359–374.

[17] Rooke, D. P., "An Improved Compounding Method for Calculating Stress Intensity Factors," *Engineering Fracture Mechanics,* Vol. 23, No. 5, 1986, pp. 783–792.

[18] Kobayashi, A. S., "A Simple Procedure for Estimating Stress Intensity Factors in Region of High Stress Gradient," *Significance of Effects in Welded Structures,* University of Tokyo, 1974, pp. 127–143.

[19] Rooke, D. P., Baratta, F. J., and Cartwright, D. J., "Simple Methods of Determining Stress Intensity Factors," *Engineering Fracture Mechanics,* Vol. 14, 1981, pp. 397–426.

[20] Wilhelm, D., FitzGerald, J., Carter, J., and Dittmer, D., "An Empirical Approach to Determining K for Surface Cracks," *Advances in Fracture Research,* Pergamon Press, New York, 1980, pp. 11–21.

[21] Liu, A. F. and Kan, H. P., "Growth of Corner Cracks at a Hole," *Journal of Engineering Materials and Technology,* Vol. 104, 1982, pp. 107–114.

[22] Shin, C. S., "The Stress Intensity of Corner Cracks Emanating from Holes," *Engineering Fracture Mechanics,* Vol. 37, No. 5, 1990, pp. 423–436.

[23] Reemsnyder, H. S., "Fatigue Life Extension of Riveted Connections," *Proceedings of American Society of Civil Engineers,* Vol. 101, No. ST12, Dec. 1975, pp. 2591–2608.

[24] Grandt, A. F., Jr. and Gallagher, J. P., "Proposed Fracture Mechanics Criteria to Select Mechanical Fasteners for Long Service Lives," *Fracture Toughness and Slow-Stable Cracking, ASTM STP 559,* American Society for Testing and Materials, West Conshohocken, PA, 1974, pp. 283–297.

[25] Petrak, G. J. and Stewart, R. P., "Retardation of Cracks Emanating from Fastener Holes," *Engineering Fracture Mechanics,* Vol. 6, 1974, pp. 275–282.

[26] Reemsnyder, H. S. and Demo, D. A., "Fatigue Cracking in Welded Crane Runway Girders: Causes and Repair Procedures," *Iron and Steel Engineer,* Vol. 5, April 1978, pp. 52–56.

[27] Heller, M., Hill, T. G., Williams, J. F., and Jones, R., "Increasing the Fatigue Life of Cracked Fastener Holes Using Bonded Repairs," *Theoretical and Applied Fracture Mechanics,* Vol. 11, 1989, pp. 1–8.

[28] Rooke, D. P. and Cartwright, D. J., *Compendium of Stress Intensity Factors,* H. M. Stationery Office, London, 1976.

[29] Tada, H., Paris, P. C., and Irwin, G. R., *The Stress Analysis of Cracks Handbook,* 2nd ed., Paris Productions Inc., St. Louis, MO, 1985.

[30] Murakami, Y., Editor-in-Chief, *Stress Intensity Factors Handbook,* Pergamon Press, New York, 1987.

[31] Blackburn, W. S., "Calculation of Stress Intensity Factors for Straight Cracks in Grooved and Ungrooved Shafts," *Engineering Fracture Mechanics,* Vol. 8, 1976, pp. 731–736.

[32] Daoud, O. E. K., Cartwright, D. J., and Carney, M., "Strain-Energy Release Rate for a Single-Edge-Cracked Circular Bar in Tension," *Journal of Strain Analysis,* Vol. 13, No. 2, 1978, pp. 83–89.

[33] Salah el din, A. S. and Lovegrove, J. M., "Stress Intensity Factors for Fatigue Cracking of Round Bars," *International Journal of Fatigue,* Vol. 3, 1981, pp. 117–123.

[34] Daoud, O. E. K. and Cartwright, D. J., "Strain Energy Release Rates for a Straight-Fronted Edge Crack in a Circular Bar Subject to Bending," *Engineering Fracture Mechanics,* Vol. 19, No. 4, 1984, pp. 701–707.

[35] Daoud, O. E. K. and Cartwright, D. J., "Strain-Energy Release Rate for a Circular-Arc Edge Crack in a Bar Under Tension or Bending," *Journal of Strain Analysis,* Vol. 20, No. 1, 1985, pp. 53–58.

[36] Nord, K. J. and Chung, T. J., "Fracture and Surface Flaws in Smooth and Threaded Round Bars," *International Journal of Fracture,* Vol. 30, No. 1, 1986, pp. 47–55.

[37] Raju, I. S. and Newman, J. C., Jr., "Stress Intensity Factors for Circumferential Surface Cracks in Pipes and Rods under Tension and Bending Loads," *Fracture Mechanics, Seventeenth Volume, ASTM STP 905,* American Society for Testing and Materials, West Conshohocken, PA, 1986, pp. 789–805.

[38] Hojfeldt, E. and Ostervig, C. B., "Fatigue Crack Propagation in Shafts with Shoulder Fillets," *Engineering Fracture Mechanics,* Vol. 25, No. 4, 1986, pp. 421–427.

[39] Toribo, J., Sanchez-Galvez, V., and Astiz, M. A., "Stress Intensification in Cracked Shank of Tightened Bolt," *Theoretical and Applied Fracture Mechanics,* Vol. 15, 1991, pp. 85–97.

[40] Toribio, J., "Stress Intensity Factor Solutions for a Cracked Bolt Loaded by a Nut," *International Journal of Fracture,* Vol. 53, 1992, pp. 367–385.

[41] Carpinteri, A., "Stress Intensity Factors for Straight-Fronted Edge Cracks in Round Bars," *Engineering Fracture Mechanics,* Vol. 42, No. 6, 1992, pp. 1035–1040.

[42] de Koning, A. U., Lof, C. J., and Schra, L., "Assessment of 3D Stress Intensity Factor Distributions for Nut Supported Threaded Rods and Bolt/Nut Assemblies," NLR CR 96692 L, National Aerospace Laboratory, (NLR), 1996.

[43] Fischer, D. F., Till, E. T., and Rammerstorfer, F. G., "Fatigue Cracks in Bolt Threads," *Fatigue of Steel and Concrete Structures,* International Association of Bridge and Structural Engineers, Eth-Honggerberg ch 8093, Zurich, 1982, pp. 725–732.

[44] Tafreshi, A. and Dover, W., "Stress Analysis of Drillstring Threaded Connections Using the Finite Element Method," *International Journal of Fatigue,* Vol. 15, No. 5, 1993, pp. 429–438.

[45] Reibaldi, G. and Eiden, M., "SpaceLab Bolt Fracture Mechanics Analysis," European Space Agency Working Paper No. 1274, March 1981.

[46] Trantina, G. G., deLorenzi, H. G., and Wilkening, W. W., "Three-Dimensional Elastic-Plastic Finite Element Analysis of Small Surface Cracks," *Engineering Fracture Mechanics,* Vol. 18, No. 5, 1983, pp. 925–938.

[47] Ng, C. K. and Fenner, D. N., "Stress Intensity Factors for an Edge Cracked Circular Bar in Tension and Bending Method," *International Journal of Fracture,* Vol. 36, 1988, pp. 291–303.

[48] Mattheck, C., Morawietz, P., and Munz, D., "Stress Intensity Factors of Sickle-Shaped Cracks in Cylindrical Bars," *International Journal of Fatigue,* Vol. 7, 1985, pp. 45–47.

[49] Caspers, M. and Mattheck, C., "Weighted Average Stress Intensity Factors of Circular-Fronted Cracks in Cylindrical Bars," *Fatigue and Fracture of Engineering Materials and Structures,* Vol. 9, No. 5, 1987, pp. 329–341.

[50] Springfield, C. W. and Jung, H. Y., "Investigation of Stress Concentration Factor—Stress Intensity Factor Interaction for Flaws in Filleted Rods," *Engineering Fracture Mechanics,* Vol. 31, No. 1, 1988, pp. 135–144.

[51] Cipolla, R. C., "Stress Intensity Factor Approximation for Cracks Located at Threaded Root Region of Fasteners," *Structural Integrity of Fasteners, ASTM STP 1236,* American Society for Testing and Materials, West Conshohocken, PA, 1995, pp. 108–125.

[52] Athanassiadis, A., Boissenot, J. M., Brevet, P., Francois, D., and Raharinaivo, A., "Linear Elastic Fracture Mechanics Computations of Cracked Cylindrical Tensioned Bodies," *International Journal of Fracture,* Vol. 17, 1981, pp. 553–566.

[53] Cipolla, R. C., "Stress Intensity Factor Approximations for Semi-Elliptical Cracks at the Thread Root of Fasteners," *Improved Technology for Critical Bolting Applications,* MPC-Vol. 26, ASME, 1986, pp. 49–58.

[54] Popov, A. A. and Ovchinnikov, A. V., "Stress Intensity Factors for Circular Cracks in Threaded Joints," *Strengths of Materials,* Vol. 15, No. 11, 1983, pp. 1586–1589.

[55] Lefort, P., "Stress Intensity Factors for a Circumferential Crack Emanating from a Notch in a Round Tensile Bar," *Engineering Fracture Mechanics,* Vol. 10, 1978, pp. 897–904.

[56] Lorentzen, T., Kjaer, N. E., and Henriksen, T. K., "The Application of Fracture Mechanics to Surface Cracks in Shafts," *Engineering Fracture Mechanics,* Vol. 23, No. 6, 1986, pp. 1005–1014.

[57] MacKay, T. L. and Alperin, B. J., "Stress Intensity Factors for Fatigue Cracking in High-Strength Bolts," *Engineering Fracture Mechanics,* Vol. 21, No. 2, 1985, pp. 391–397.

[58] Forman, R. G. and Shivakumar, V., "Growth Behavior of Surface Cracks in the Circumferential Plane of Solid and Hollow Cylinders," *Fracture Mechanics, Seventeenth Volume, ASTM STP 905,* American Society for Testing and Materials, West Conshohocken, PA, 1986, pp. 59–74.

[59] Bush, A. J., "Experimentally Determined Stress Intensity Factors for Single-Edge-Crack Round Bars Loaded in Bending," *Experimental Mechanics,* Vol. 16, No. 6, 1976, pp. 249–280.

[60] Ouchterlony, F., "Extension of the Compliance and Stress Intensity Formulas for the Single Edge Crack Round Bar in Bending," *Fracture Mechanics Methods for Ceramics, Rocks, and Concrete, ASTM STP 905,* American Society for Testing and Materials, West Conshohocken, PA, 1981, pp. 237–256.

[61] Bush, A. J., "Stress Intensity Factors for Single-Edge-Crack Solid and Hollow Round Bars Loaded in Tension," *Journal of Testing and Evaluation,* Vol. 9, No. 4, 1981, pp. 216–223.

[62] Makhutov, M., Zatsarinny, V., and Kagan, V., "Initiation and Propagation Mechanics of Low Cycle Fatigue Cracks in Bolts," *Advances in Fracture Research (Fracture 81),* D. Francois, Ed., Pergamon Press, Oxford, UK, Vol. 2, 1982, pp. 605–612.

[63] Daoud, O. E. K. and Cartwright, D. J., "Stress Intensity Factors and Strain-Energy Release Rates for Edge-Cracked Circular Bars," *The Mechanism of Fractures,* V. S. Goel, Ed., American Society for Metals, Metals Park, OH, 1986, pp. 223–229.

[64] James, L. A. and Mills, W. J., "Review and Synthesis of Stress Intensity Factor Solutions Applicable to Cracks in Bolts," *Engineering Fracture Mechanics,* Vol. 30, No. 5, 1988, pp. 641–654.

[65] Liu, A. F., "Evaluation of Current Analytical Methods for Crack Growth in a Bolt," *Proceedings of the 17th ICAF Symposium,* Stockholm, May 1993.

[66] Liu, A. F., "Behavior of Fatigue Cracks in a Tension Bolt," *Structural Integrity of Fasteners, ASTM STP 1236,* American Society for Testing and Materials, West Conshohocken, PA, 1995, pp. 126–140.

[67] Sih, G. C., *Handbook of Stress Intensity Factors,* Institute of Fracture and Solid Mechanics, Lehigh University, Bethlehem, PA, 1973.

[68] Gangloff, R. P., "Quantitative Measurements of the Growth Kinetics of Small Fatigue Cracks in 10Ni Steel," *Fatigue Crack Growth Measurement and Data Analysis, ASTM STP 738,* American Society for Testing and Materials, West Conshohocken, PA, 1981, pp. 120–138.

[69] Coles, A. S., Johnson, R. E., and Popp, H. G., "Utility of Surface-Flawed Tensile Bars in Cyclic Life Studies," *Journal of Engineering Materials and Technology,* Vol. 98, 1976, pp. 305–315.

[70] Dover, W. D., "Crack Shape Evolution Studies in Threaded Connections Using A.C.F.M.," *Fatigue of Engineering Materials and Structures,* Vol. 5, No. 4, 1982, pp. 349–353.

[71] Michael, D. H., Collins, R., and Dover, W. D., "Detection and Measurement of Cracks in Threaded Bolts with an A.C. Potential Difference Method," *Proceedings of the Royal Society,* Vol. A385, 1983, pp. 145–168.